# FOREWARNED

# PAUL GOODWIN

# FOREWARNED

—

## A SCEPTIC'S GUIDE TO PREDICTION

Biteback Publishing

First published in Great Britain in 2017 by
Biteback Publishing Ltd
Westminster Tower
3 Albert Embankment
London SE1 7SP
Copyright © Paul Goodwin 2017

ISBN 978-1-78590-222-2

A CIP catalogue record for this book is available from the British Library.

Set in Minion Pro

Printed and bound in Great Britain by
CPI Group (UK) Ltd, Croydon CR0 4YY

*To my wife, Chris*

# CONTENTS

# INTRODUCTION

**M**y hairdresser wondered if they were pumping something into the atmosphere – whoever 'they' were.

'Every time I turn on the TV news,' he said, 'there's another shock. The thing you least thought would happen, happens. Recently, everything seems to be unpredictable.'

We were speaking a few days after Brexit, Britain's momentous decision to leave the European Union after more than forty years. In the hours before the first results were declared, the British pound had surged on the international markets in the anticipation of a vote to 'remain'. Even Nigel Farage, the beer and nicotine-loving long-term enemy of the EU, had conceded: 'The Remain side have edged it.' Yet by the morning Farage was celebrating that his life's work was complete and calling for 23 June to be declared Britain's Independence Day.

It wasn't just the Brexit vote, the hairdresser pointed out. Leicester City, an inexpensive team of journeymen, had recently broken the boring oligopoly of wealthy soccer clubs who'd assumed joint ownership of England's Premiership league title. Leicester had won the championship as 5,000-to-1 outsiders. And in the European Championships, England, a team of millionaires, had been kicked out of the competition by tiny Iceland, coached by a part-time dentist.

Then there was the rise and rise of Donald Trump towards the possible presidency of the US. Trump was initially seen as a joke candidate; each outrageous utterance he made should surely in ordinary times have resulted in his political suicide. Instead, buoyed by sensational publicity, he became the presumptive candidate for the Republican Party. Of course, the situation only escalated when Trump defeated Hillary Clinton to become the 45th President. A £5 triple bet on Leicester's triumph, Brexit and Trump's victory would have won you £15 million – Paddy Power put the odds at 3 million to 1!

It was as if the 'normal' laws of predictability were breaking down, as if physical certainties like gravity or time had started to behave in odd ways. Who would have thought it? Nationalists wiped the Labour Party off the political map in Scotland, Labour's most rebellious Member of Parliament, Jeremy Corbyn, was crowned the party's leader and a host of predictably 'nice' former TV celebrities were jailed for sexual crimes. And, in what seemed like the recent past, bastions of apparent commercial stability like Lehman Brothers and Woolworths crashed like mountains into an ocean of economic turmoil.

Perhaps it was always like this: today's shock is tomorrow's norm. The outbreak of two world wars in the twentieth century now seems unsurprising given the rationalisation of hindsight, yet they must have seemed incredibly shocking at the time. A quarter of a century on, the rapid, and relatively peaceful, dissolution of the Soviet Union and the end of the Cold War seem to have been inevitable, but then I remember the stomach-churning fear I felt as a child when the Cuban Missile Crisis dominated the news: we were told it would take only four minutes for everything to come to an end. During the turbulent '60s, our world's fragile existence seemed dependent on the whims and enigmatic calculations of distant politicians.

## The besieged forecaster

Whether the world is becoming more unpredictable or not, these shocks are great for TV newshounds and audience ratings. But they reflect badly on the forecasters who are supposed to predict what is going to happen. I was careful not to let it slip to the hairdresser that I'd worked in forecasting for years. I recalled all too well the comments of a US immigration official who asked me why I was visiting America.

'To speak at a forecasting conference,' I replied, expecting him to be impressed.

'What are yer forecasting – the weather?'

'No, my talk's about sales forecasting.'

'Oh I used to do that. It's a load o' trash. Yer may as well toss a coin, mightn't you?'

Jet-lagged, and concerned that he might not let me into the country, I agreed with alacrity and inserted a false laugh to signify we were both sharing the same mischievous secret. Looking back, he probably saw me as some sort of con man on a junket bound for America to spout rubbish dressed up as science.

Being attacked in this way goes with the territory of being a forecaster. 'Never trust an economic forecast', advised the headline of an article by the *Financial Times* columnist Tim Harford, before listing a series of apparently grossly over-optimistic or over-pessimistic forecasts for the UK economy since 1995. 'In the Eurozone', he added, 'forecasting over the past few years has been so wayward that it is kindest to say no more'.[1]

'Whoops: economic forecast wrong within weeks', announced a headline in the Huffington Post arguing that the USA's Congressional Budget Office forecasts were 'no better than wild guesses'.[2] Similar sentiments led the *Sunday Times* columnist Dominic Lawson to write in December 2014: 'My year-end forecast: there is no future for prediction'.[3] Lawson was mainly reacting to the failure of economic forecasters to anticipate the 40 per cent drop in world oil prices that had occurred since June of that year.

Even the Queen hinted at her royal disapproval of economists for their failure to predict the catastrophic events of 2008, when the entire banking system in many countries teetered on the brink of total collapse. While visiting the London School of Economics to open a new building she was given a briefing on the causes and effects of the credit crisis. 'Why did nobody notice it? If these things were so large how come everyone missed it?' she asked.

In Britain the popular BBC weatherman Michael Fish was never allowed to forget his forecast of October 1987 when he said: 'Earlier on today, apparently, a woman rang the BBC and said she'd heard there was a hurricane on the way. Well, if you're watching, don't worry, there isn't.' By the following morning people in the south-east of England were recovering from the worst storm in 300 years. It caused widespread damage to buildings, brought down thousands of trees and killed at least nineteen people.

It was a pity for Fish. Looking like a convivial university don, he had a penchant for making appearances on light entertainment shows, and had joined the Met Office in 1962. By the time he made his fateful broadcast he'd had a quarter of a century in the forecasting business. A colleague described him as 'the last of the true weathermen [who] can actually interpret the skies – he can do the weather forecast the hard way … nowadays most of the decisions are made by the computer'. In his time he must have been involved with thousands of forecasts and the vast majority of these were probably accurate. But most people will only recall the rare forecasts that are regarded as spectacular failures. Get a thousand forecasts right but one very wrong and you'll probably be judged by that single forecast. In forecasting, reputations are hard won, but very easily lost.

In this case, perhaps not totally lost. Now Fish appears to revel in the attention that his famous broadcast attracted. A video of it appears on his website, where he's described as a national treasure. You can even buy Michael Fish 'Retro Weather' fridge magnets, as well as Michael Fish 'Weather Changing' mugs.

Some forecasters suffer more than a mere blasting from newspaper columnists or royalty – they get sent to jail.

The central Italian city of L'Aquila lies in a region where there are major faults in the earth's crust and is near where the African and Eurasian tectonic plates are colliding. It saw at least seven earthquakes between 1315 and 1706. The city is also built on the bed of an ancient lake, resulting in a soil structure that exacerbates the effect of any seismic activity. In early 2009 the region around L'Aquila began to be affected by minor earthquakes, causing anxiety for local people who wanted to know what these tremors portended.

On 30 March there was a quake that was severe enough to persuade some people to sleep outdoors. The Civil Protection Department immediately called a meeting with the National Commission for Forecasting and Preventing Major Risks to establish the risk that a major earthquake was imminent. The meeting was attended by some of the leading experts in the field – seismologists, a volcanologist, a geophysicist and seismic engineers – together with the deputy head of the Civil Protection Department, Bernardo de Bernardinis.

At the meeting, which lasted for an hour and was widely

reported in the media, the experts stated that with the current scientific understanding of earthquakes it would be impossible to predict if, and when, a strong quake would occur, but that a major tremor, like the one that destroyed the city in 1703, was 'unlikely'. The recent seismic activity, they argued, was not necessarily a sign that a stronger shock was on its way.

Reassured, L'Aquila's mayor and de Bernardinis went straight to a press conference where they conveyed the message that earthquakes were impossible to predict, but that scientists did not expect a major one. In a separate broadcast interview, de Bernardinis argued that the experts had indicated that the recurring quakes did not pose any danger. Indeed, as the activity involved a discharging of energy from the earth's crust, it should instead be seen as favourable. The public should stop worrying and enjoy a glass of wine instead.

At 3.32 a.m. on 6 April a huge earthquake (rated 6.3 on the Richter scale) destroyed much of the city, killing 309 people and injuring over 1,500. More than 90 per cent of the population was left homeless.

But, for the scientific community, a shock of a different kind was on the way. In September 2011, the seven experts who had participated in the fateful meeting of the National Commission in 2009 appeared before a court charged with the criminal manslaughter of twenty-nine inhabitants. These were people who had chosen to stay indoors on the night of the disaster. It was argued that the scientists had failed in their 'institutional duty' to assess and communicate risk. It was claimed that their reassuring words had prevented people from following their long-standing habit of staying outdoors after smaller shocks had been experienced. After a trial that lasted over a year, all seven defendants were found guilty and sentenced to six years in prison. They were also ordered to pay more than €9 million to compensate survivors of the disaster.

Many scientists expressed outrage at the verdicts. Over 5,000 of them signed an open letter to the Italian President in support of the experts. Some argued that science itself had been put on trial and drew parallels with the treatment of Galileo by the Catholic Church in the seventeenth century. Lord May, a former president of the Royal Society, said the verdicts were 'truly shocking and revealed appalling ignorance of the basic nature of scientific enquiry'. The verdicts were overturned on appeal in November 2014, but the commission members had already spent time in jail and there were cries of 'shame' in the courtroom from earthquake survivors when the successful appeal was announced.

## A safer occupation?

For those who want a quieter life, long-term forecasting can offer a safe haven. Forecasts of what the world will be like in twenty-five or fifty years' time are likely to be much less reliable than those made for the short-term but, with luck, these forecasts will be long forgotten before future generations have had a chance to mock them. Even if they are remembered, the forecaster might not even be around to witness the public's gleeful disdain. Usually, long-range predictions tell us more about the way the world was when the forecast was made than they do about the future. It can be great fun looking back at these predictions with the smugness of knowing what really happened.

In 1981, David Wallechinsky, Amy Wallace and Irving Wallace published *The Book of Predictions*. They had sought out and asked the 'leading minds on Earth' for their predictions of what was to happen in the years to come. These experts predicted that by the year 2000, 50,000 people would be living and working in space. The first humans would land on Mars in 2010 and the first permanent colony on the moon would open by 2015. By 1991 petrol-powered cars would be banned from metropolitan areas in the US, Europe and Japan, while, by 2000, all Persian Gulf countries would run out of oil. International terrorists would use nuclear weapons to destroy a major world capital in 2010, but on a happier note, by 1992 there would be cures for cancer, the common cold and other viral ailments. Indeed, earthly immortality – even circumventing accidental death – would be achieved by 2060.

It's fair to say that, of the thousands of predictions in the book, some have inevitably been realised. For example a Catholic priest, who was also a professor of sociology, predicted the end of the communist government in the USSR before 1990 and the independence of the Ukraine and Soviet colonies in Eastern Europe. Nevertheless, another expert predicted that the USSR would draw level with the West in the sophistication of its military electronics by 1993. A former CIA analyst went even further, predicting that by 1993 the US would cease to be a great power and the Soviet Union would rule almost the entire world. We could go on.

The brilliant Lord Kelvin, president of the Royal Society, who died in 1907, told the world that X-rays would prove to be a hoax and radio had no future. In *A Short History of the Future*, published in 1936, journalist John Langdon-Davies predicted that by 1960 work would be limited to just three hours a day and crime would cease to exist by 2000. Tom Watson, chairman of IBM, famously announced in 1943: 'I think there is a world market for maybe five computers.' Decca Records rejected The Beatles in 1962 because 'we don't like their sound, and guitar music is on the way out'.

### The pain in our brain

Predictions that are wide of the mark like these tempt us to see forecasters in the same comical light as those doomed early pioneers of flying we sometimes see on flickering monochrome footage – optimistically flapping their bird-like wings before crashing off the ends of piers. Or we may see them as the deluded modern versions of ancient astrologers and necromancers, dabbling with strange mathematical symbols and scientific babble rather than star signs or dead bodies. Trying to forecast the future – we may think, surely it's futile… and then a pundit appears on television to tell us what the economic outlook is, or who will win an election, and we sit on the edge of our seats, absorbing every word. How can we be so ambivalent?

It's all to do with the way our brains handle uncertainty. Most of us hate it. Some psychologists, such as David Rock, have argued that we hanker after certainty in the same way that we crave primary rewards such as food and sex. He suggests that the brain treats uncertainty like a type of pain. It's associated with the uncomfortable feeling that we can't control the future. When those we perceive to be experts tell us what is going to happen, we are offered reassurance that helps to relieve this 'pain'. Confident pundits on television or in the newspapers who tell us that the stock market will soon start to rise, or extrovert TV weather forecasters, backed up by their supercomputers and satellites, are the perfect analgesics.

But when the smooth self-assurance of the pundits or twenty-first century advanced science and technology appear to have misled us, we feel cheated: what do we pay these people for? Perhaps we are now so enamoured with experts and technology that we subconsciously believe they should be able to predict the future exactly? After all, the UK's Meteorological Office has some of the best technology in the world at its disposal. In 2014 it announced plans on its website for a £97 million supercomputer that will perform 16,000 trillion calculations per second, will weigh the equivalent of eleven double-decker buses and have 120,000 times more memory capacity than a market-leading smartphone. In a world where we think we have largely tamed nature, doubled our life expectancy, invented the internet and landed scientific rovers on Mars, we are perhaps annoyed that the future – even the short-term future – remains a wild and untamed place. But, as we'll see later, our annoyance may only be temporary. As our memories fade we may start reaching for the analgesics again. Reassurance is a powerful drug.

While we are all consumers of forecasts – from weather forecasts to stock market forecasts, from election forecasts to sports forecasts – we are also all forecasters ourselves. We couldn't make decisions if we weren't. Should we choose that January holiday in the Canary Islands? We need to predict what it will be like there and how enjoyable it might be. Should we change jobs? We need to predict whether the new job will be better than our current one. Some of us go further and try to predict who will win a sports event or which party will win an election in the hope of gaining a big payout. But can we trust our own predictive skills?

My aim in this book is to show you when you can trust a forecast – be it from an expert or one we make ourselves – and when it's best to take it with a pinch of sodium chloride. It's a sceptical consumer's guide to prediction – but, of course, when you forecast your own future you may be both producer and consumer. You won't find much about astrology or necromancy here – I know very little about these. But I will tell you what scientific research reveals about our ability to calculate and miscalculate the future. Despite the hairdresser's fears that a secretive world elite may be conspiring to distort the hidden laws of predictability, as well as Dominic Lawson's diatribe, we'll see that under the right conditions forecasts can be highly reliable. We'll encounter some amazing examples of where ignorance and lack of expertise can make us brilliant forecasters. We'll find that computers crunching through mountains of data can make surprisingly accurate discoveries and predictions about the quirks in our behaviour. We'll even see that forecasts that are presented honestly and realistically can enable us to outsmart the risks of tomorrow's world.

But we'll also explore the downsides of forecasting. We'll see that it's often contaminated by self-interest and politics – even respected world institutions have been caught out distorting their forecasts to suit their political ends. We'll find out when our confident predictions, based on our own judgement, will be predictably wrong. We'll meet the researchers who make big claims for their super-mathematical forecasting methods but omit to test their ability to forecast. And we'll see how our dismissal of the past as a different, irrelevant world and our proclivity for believing that each new situation is unique can cost us dearly – very dearly.

Of course, in some situations, if we are honest, we should admit that the future is completely unpredictable. In the penultimate chapter we'll look at how we can attempt to position ourselves so that we'll survive, and even thrive, whatever the future throws at us. We'll also look at the diametrically opposite situation – where we can predict the future with high accuracy but it's perhaps best not to know. Making a forecast can change the world and our futures – and not always for the better.

Our predictions of the future are inherently bound up with psychology, history, politics, sociology, statistical analysis and computer power. It's a fascinating combination but it means that discerning which forecasts to believe and which to dismiss can be difficult. The following chapters should make this easier. We will start by looking at examples of forecasts based on split-second judgements that have turned out to be highly accurate. As we'll see, our intuition can have surprising powers of foresight as long as the conditions are right.

# CHAPTER 1

# NEURONS GALORE

### Thin slices of brilliance

It had been an astonishing night. For months the experts had predicted that Britain's 2015 general election was heading for a hung parliament, with no single political party able to govern on its own. The politicians had queued up with their 'red lines' – positions they said they would not compromise on if invited to join a coalition. The BBC's website had a game called 'Can you build a majority?' that invited people to combine the number of seats that might be won by different parties.[1] The aim was to see if they could reach the magic figure of 326 seats that would produce a majority. Some people worried about weeks of horse-trading before a weak, unstable government dared to see if it could survive a hostile parliament; others predicted that there was bound to be a second election that year.

It seemed so certain – day after day the pollsters told us that support for the two main parties was tied at around 34 per cent. It was getting boring. Whatever the politicians said or promised, the polls hardly moved. Nate Silver, billed as 'the rock star statistician' and the man who correctly called every state in the 2012 US election, had an entire episode of the leading BBC current affairs show *Panorama*, devoted to his forecasting methods. He toured the country in an American caravan (or at least the programme gave that impression), playing bingo with northerners and eating Cornish pasties in the south before declaring there could be an 'an incredibly messy outcome'. The Conservatives, he predicted, would win 283 seats and Labour 270. One unnamed gambler placed over £200,000 on there being a hung parliament at odds of 2/9.[2]

Then we found ourselves on the morning of 8 May, exhausted but shocked after a sleepless night of swings, shares of the vote and dazzling graphics on television. The Conservatives were heading for a clear majority with their 330 seats, while Labour lagged behind with only 232. It didn't seem possible. How could all of the data collected by armies of market researchers, the sophisticated analysis, the computing power and the expertise have led to such a misleading prediction? Michael Lewis-Beck, a professor at the University of Iowa and well-known election forecaster (and accomplished poet and novelist to boot) told me he'd been involved in a big London conference on forecasting UK elections two months earlier. 'No one forecast a Conservative majority,' he told me. 'It's all a big splash, and a mess.'

Theories about the debacle were soon hitting the airwaves. Shy Conservatives had misled the pollsters, embarrassed to admit they were voting for politicians who belonged to 'the nasty party'. English voters had changed their minds at the last minute, frightened that a minority Labour government would be beholden to a surging Scottish National Party. It was tempting to question whether human behaviour, at least when examined en masse, could be reliably predicted through so-called scientific methods. Computer algorithms that effortlessly process gigabytes of data at unimaginable speeds might be perfect at landing a spacecraft on a comet after a journey of 6.4 billion kilometres, or at predicting when there will next be a total eclipse of the sun visible from any specified spot on Earth. But perhaps the vagaries of humans posed a challenge of a different order.

After all, computers are dumb. They rely on blind sets of formulae that are usually gross simplifications of a complex world they have no true understanding of. Ask a computer to interpret a phrase such as 'time flies like an arrow' and it might think that there is a species of insect called 'time flies' that feed on arrows. Ask it to forecast a company's sales and, as it crunches through masses of past data searching for trends, it will be ignorant of a customer's subtle body language or tone of voice and what they reveal; that despite his or her smiles the customer is tiring of a product and probably looking at alternatives. A sales representative or accounts manager might intuitively register these cues. As a manager in a pharmaceutical company told me, while referring to the company's expensive sales forecasting software, 'it's there, it's useful, but it needs to be managed since no way can it have the market intelligence'. The manager's comments would have been echoed by executives at Nike, in 2001. Having relied unquestioningly on the automated sales forecast of its computers, the company lost millions in sales and a third of its value on the stock market. Nike had ordered $90 million worth of shoes, such as the Air Garnett 2, that turned out to have very disappointing sales. Worse still, it hadn't ordered enough of popular styles such as the Air Force 1, losing out on $90 million to $100 million of possible sales.[3]

So can predictions based on human judgement do any better? Our brains are marvellous structures. They are the most complex systems in the known universe and have around 100 billion neurons that are interconnected through trillions of synapses. With all this processing power in our heads, much of it unconscious, as well as our rich experience of the world and the intricacies of human existence, surely we can make more accurate predictions of human behaviour than unthinking computers?

There is mounting evidence that this might be true, particularly with settings that favour us. Indeed, there could be a much easier way of obtaining reliable election predictions. Simply flash photographs of the candidates in front of someone who knows nothing about the election and has no idea who the candidates are. Ask them instantly to rate the competence of the person in the picture and assume the highest-rated candidate will be the most likely winner.

Alexander Todorov of Princeton University and his colleagues did just that. They showed so-called naïve participants pairs of black-and-white headshot photographs of candidates in the US Senate and House elections between 2000 and 2004.[4] The participants only saw the photographs for one second and were then asked to judge which candidate was the most competent. If a face was recognised then the judgement was discarded, so only facial features could be used in the assessment. The candidate who was judged to be the most competent won in over 71 per cent of the Senate elections, and in over 66 per cent of the House elections – results that were much better than expected, given that they were based on chance guesswork. Scott Armstrong of the Wharton School in Pennsylvania and his fellow researchers found similar results when they asked New Zealand schoolgirls to rate the competence of candidates in US presidential primary contests, again based solely on their photographs.[5] And, in a more recent study, individuals in the US made reliable predictions of how eighteen candidates would fare in Bulgaria's 2011 presidential elections.[6]

These results suggest that voters may judge election candidates based on intuitive impressions of their competence, irrespective of their track records or policies or even their attractiveness. But they also show the potential power of harnessing intuition – that effortless, unconscious ability to process complex information and make instant judgements.

There are plenty of other surprising indications that predictions based on these judgements can be uncannily accurate. For example, Harvard University researchers, Nalini Ambady and Robert Rosenthal,[7] found that, after watching silent video clips of teachers in front of their classes, lasting between two and ten seconds, complete strangers could accurately predict what students' ratings of their teachers would be at the end of the semester. The researchers referred to the short excerpts of behaviour that people observed as 'thin slices' and remarked on the wealth of information that they conveyed.

Other studies have provided consistent findings. Ratings of managers, based on only ten-second clips of videoed interviews, provided reliable predictions of how their supervisors viewed their job performance.[8] 'Naïve' judges were able to predict the effectiveness of sales people after hearing three twenty-second audio recordings of them being interviewed. Thin slices have even enabled people to predict how satisfied patients will be with their doctors[9] or assess whether a surgeon has been sued for malpractice. After hearing twenty-second clips of surgeons' voices, recorded when they were interacting with their patients, listeners rated them on a scale designed to reflect the personality trait of dominance. Those scoring higher were more likely to have been sued in the past.[10]

But significantly, all of these examples of amazing accuracy relate to predicting the behaviour of people. As social beings we have plenty of experience of how others behave. This has fed a rich biological database of past cases and allowed us to have lots of practice in predicting what our fellow beings will do, plus lots of feedback on the success or otherwise of these predictions, which enables us to learn. As with repeated practice in a sport or with a musical instrument, pathways become hardwired in our brains. This means we can instantly match a new situation to past ones and automatically determine the most likely correct response. The top-class tennis player does not have time to think carefully about how to respond to the 100 miles-per-hour serve. The response is instantly determined. The chess grandmaster, who plays multiple opponents simultaneously and beats them all, instantly recognises the patterns of pieces on each board from a vast store of past games. Without conscious thought, he or she determines the best move.[11] Intuitive predictions therefore seem to work well when we have plenty of practice, experience and feedback.

## Ignorance is power

The people whose judgements led to accurate election predictions, based on quick flashes of candidates' photographs, knew nothing about those standing for election or the policies they were espousing. If they had been provided with that information it's possible it might have distracted them from being able to predict the results. We find rating someone's competence in a split second easy. Handling lots of detailed information is a different matter, so to make things easier we often focus on just a few items of salient information, ignoring the rest. The detail we pick out can sometimes mislead us in our predictions. When listening to politicians, we recall a few unrepresentative soundbites we happened to hear on the radio. Or we remember liking what a candidate said about combatting global warming, but we miss what she said about raising taxes. In prediction, a little knowledge can be dangerous – sometimes ignorance can be an advantage.

There is another way in which ignorance can be beneficial – when we make predictions based simply on what we recognise.[12] Dilek Önkal, of Bilkent University in Ankara, and Peter Ayton of City University in London cunningly spotted an opportunity to test this possibility a few years ago.[13] They pitted the predictive abilities of keen, British football fans against those of Turkish students who knew little about the English game. Both groups had to forecast the winners of matches in the third round of the 1994 FA Cup, where top teams can be drawn randomly against lower-league sides. Dilek told me that many of the Turkish students protested it was an unfair contest: how were they expected to make accurate predictions? Yet, surprisingly, their performance was almost as good as the British participants – they correctly forecasted the outcome of 62.5 per cent of games compared to the British score of 65.6 per cent. How did they do it? They apparently used a simple rule. If they recognised the name of one team and not the other, they forecasted that the team they recognised would win. The names of bigger teams are more likely to be recognised – even people not interested in football have probably heard of Manchester United – and because they are bigger they are also more likely to win, so the rule worked well. The British fans, who presumably concentrated on details like the teams' recent form, what the teams' coaches had to say before the game and which players were likely to be selected, could do little better. This has been referred to as the 'less is more effect'.

Benjamin Scheibehenne and Arndt Bröder of the Max Planck Institute even found that people relying on recognition to predict winners in the 2005 Wimbledon Championships were as good as, or outperformed, predictions based on official ATP (Association of Tennis Professionals) rankings and the seedings of Wimbledon experts.[14] Amateur players and laypeople were asked to show players' names they recognised and forecast who would win matches. Overall, they correctly predicted the winner in 70 per cent of the matches. The researchers suggested that this was likely to be due to the mass media paying more attention to stronger players, so non-experts were more likely to recognise their names.[15]

Of course, recognition is not always guaranteed to work. The British ski jumper Eddie 'The Eagle' Edwards became famous precisely because of his lack of success. He finished last in the 70-metre and 90-metre events at the 1988 Winter Olympics in Calgary, but was courted by the media as an example of an underdog bravely striving for success without backing from funders or top coaches. Also, some players remain famous long after the peak of their success. For simple recognition to work as an accurate predictor there must be a correlation between how recognisable a possible outcome is and the probability that the outcome will occur – the tennis player or football team that is recognised needs to have the highest probability of winning. If both teams or players in a game are recognised, because a person has lots of knowledge about the sport, then clearly recognition cannot be used to make the prediction. The success of the strategy relies on ignorance.

## More on less is more

Recognition involves little or no mental effort and it does best when we know less. But what if lots of information is foisted on us before we make our predictions? For example, a manager might have to predict which of two new products, the Alpha and the Zeta, will have the highest sales over the next year. She has information on the amounts that will be spent advertising the products in newspapers and on television, the percentage of people who indicated during market research that they will 'definitely buy' each product, the price that will be charged for the products, whether press reviews of the products were positive or negative and so on.

Common sense suggests that the more information the manager has, the more accurate her prediction will be. But we'll be asking a lot of her. There are limits to the amount of information we can handle at once and she will have to rely on her memory to recall how these factors influenced sales of earlier products. In addition, she'll have the difficult task of balancing factors that favour one product over the other against those that give it a disadvantage. Computers excel in processing large amounts of data and their memories are less fallible than those of humans. So perhaps the manager would be better turning to computer algorithms to get her prediction. The computer would analyse data on previous products the company had launched looking for the combinations of factors that were most closely correlated with sales. It looks like there's no contest. Yet, surprisingly, large amounts of information processed by computers can, in some circumstances, lead to less accurate predictions than human judgements based on very limited information.[16]

Other researchers at the Max Planck Institute for Human Development in Berlin simulated the simple way people often make predictions. They called this strategy 'Take the Best'. If the manager used 'Take the Best' she might ask herself: which factor is most important in determining the sales of a product? She concludes this is the product's television advertising budget. But both products have a similar budget so this is unhelpful in predicting which will have the highest sales. She then considers the second most important factor, which she thinks is the percentage of people who said they would definitely buy the product. Although many people say they'll buy the product to please the market researcher and then change their minds, this still has some predictive power. Twenty per cent of people said they would definitely buy the Alpha but only 9 per cent confirmed they would buy the Zeta, so she predicts the Alpha will have the highest sales. It's as simple as that. In this case, only two factors – TV advertising budget and market research results – were considered. When researchers simulated the use of this simple strategy to make predictions in a wide range of areas, including high school dropout rates, car accident rates, fish fertility and rainfall, it tended to outperform predictions based on computer algorithms. This was despite 'Take the Best' basing its predictions on only 2.4 factors on average, while the computer algorithm used an average of 7.7.

The relative success of this simple strategy appeared to be due to the computer's tendency to seek patterns and relationships in every detail of the large sets of past data they analysed. But many of these details and apparent relationships were random quirks that were unlikely to be repeated in the future. This meant that the elaborate formulae the computers had used to produce their forecasts were excellent representations of what had happened in the past, but poor indications of what would happen in the future. Forecasters call this phenomenon 'overfitting'.

You can even allow a formula to have a closer fit to past data by including nonsensical factors. In a past election year, when inflation was 4 per cent, unemployment was 1.2 million and growth was 2 per cent, our formula might tell us that an incumbent government received 45 per cent of the votes. When we see that the government only received 32 per cent of the votes, we are disappointed by our formula's accuracy. To try to improve on this, we expand the range of information it assesses to include the annual number of umbrellas left behind on the Tube during election years. It now tells us that the government will have won 31 per cent of the votes – a massive improvement in accuracy. For a moment we get a warm feeling – thinking we've discovered some deep insight into voter behaviour. But again, we'll probably be disappointed when we test our formula's ability to forecast future elections. It has just picked up on a coincidental short-term association between voting behaviour and lost umbrellas which is unlikely to prevail in the future. Simplicity and common sense are the antidotes to overfitting.[17]

## Natural forecasting

Sometimes we use our judgement to forecast the probability of future events occurring. We might say that there's a 70 per cent chance our favourite soccer team will win the league this season, or argue that there's only a 50/50 chance that the current government will survive until the end of the year. When making these sorts of judgements we seem to be most accurate when we use what psychologists call natural frequencies,[18] rather than probabilities. Research suggests that, as we go through life, we naturally observe and record the frequencies of occurrence of events that have some importance to us. As a result we seem to be quite accurate when asked to estimate these frequencies. When people were asked in one study to estimate the relative number of restaurants operated by fast food chains, their estimates were close to the true numbers.[19]

According to Gerd Gigerenzer, also of the Max Planck

Institute, our natural facility for handling frequencies means we feel more at home with statements like: 'it snows on four days out of every 100 typical days in January' as opposed to 'there is a 0.04 probability of it snowing on a typical January day'. To see how thinking in terms of frequencies can help, let's consider the following question:

John has a vegetable stall in the open-air market that takes place every Wednesday in the town of Greenville. He operates on narrow margins and makes a loss 20 per cent of the time. Customers are often deterred by bad weather and heavy rain is expected during tomorrow's market. Which of the following is more probable?

(a) John will make a loss tomorrow.

(b) John will make a loss because of heavy rain.

Research[20] suggests that many of us would opt for b. After all, it gives us a reason for John making a loss and it's easy to envisage him returning home at the end of the day from a deserted, rain-splashed market with much of his stock unsold. Now consider the same question framed slightly differently:

John has a vegetable stall in the open-air market that takes place every Wednesday in the town of Greenville. He operates on narrow margins and makes a loss 20 per cent of the time. Customers are often deterred by bad weather and heavy rain is expected during tomorrow's market. Now consider 100 typical weeks.

Estimate how many of these weeks will see John making a loss.

Estimate how many of these weeks will see John making a loss because of heavy rain.

Clearly the answer to (a) is twenty. We don't have enough information to come up with a reliable estimate for (b), but we see it that it cannot be more than twenty and is likely to be less. John probably makes a loss due to a variety of reasons in addition to the weather. In some weeks his vegetables may be of poor quality, so even on a sunny day no one buys them. In other weeks many of his customers may be on holiday or tempted away by supermarket promotions. John making a loss specifically because of heavy rain is less likely than John making a loss for any reason, so in the first version of the question (a) is the correct answer. Many people get this wrong. Yet, when asked to think in terms of frequencies, as in the second version, most correctly judge that (a) is more likely than (b).

There has been some debate about why thinking in terms of frequencies makes it easier to make accurate probability forecasts. One explanation is that it makes the mental calculations easier. Another theory is that over thousands of years we evolved the ability to use natural frequencies to assess risks in order to survive.[21] Perhaps we noticed that in hot weather a dangerous animal would appear more frequently than in wet weather; or that rain occurred more frequently on days when there was a red sky at dawn but less frequently when there had been a red sunset the day before. Yet the idea of probability only appeared relatively recently in human history – its first occurrence appears to be in the mid-1600s, when mathematicians set out to resolve a gamblers' dispute concerning a popular dice game.[22] Our brains have therefore had less time to adapt to its demands.

### The case for intuition?

As creatures trying to survive in a hostile and challenging world, we needed to develop the ability to make accurate predictions – predictions about the behaviour of our fellow beings, predictions about what was safe and unsafe and predictions about sources of food. Evolution endowed us with a marvellous ability to make such predictions. In an instant we could form a judgement about whether another human was trustworthy or competent as a leader. If we recognised the taste of a berry then it was probably safe to eat. We knew that a prey animal was more frequently seen on the open plains when the weather was dry than when it was wet.

Our brains use up a lot of energy resources – up to 20 per cent of the calories we consume and more than any other organ.[23] There is some evidence that hard thinking taxes these resources.[24] It seems that we learned to conserve energy by making accurate predictions effortlessly and by processing small amounts of information. As we've seen, these quick predictions can beat those obtained from the most sophisticated analyses.

So should all forecasts and predictions be based on judgement and intuition? After all, the evidence is that most sales forecasts made by companies are based on managers' judgements.[25] Even weather forecasts are ultimately based on judgement, though these judgements are supported by massive computer power. Every day, thousands of people seem happy to use their judgements to bet on the results of horse races or football games. Unfortunately, as we'll see in the next chapter, in the wrong circumstances our intuition can let us down badly. In fact, we can be awful forecasters, but at the same time convince ourselves we are brilliant.

# CHAPTER 2

# A JUDGEMENT ON JUDGEMENT

**Patterns that aren't there**

Obesity is a major problem in many countries. In the USA more than two thirds of adults are considered to be overweight and obese. In the UK, between 1993 and 2013, there was almost a doubling of the proportion of men who were obese, and more than a 7 per cent increase in the proportion of women.[1] One explanation is that we have evolved to cope with a world where periods of abundance were followed by periods of famine. We needed to gorge ourselves when there was a surplus of food in order to survive the lean periods. But the problem begins when a good supply of food is constantly available, because we continue to stuff ourselves and grow fatter as food is never scarce. Our bodies aren't designed for this modern environment. Some people argue that the same applies to our minds.

Those sharp intuitive skills, which we honed in order to predict when we might be in danger from wild animals or where a harvest of luscious fruit might be found, don't always equip us for making accurate predictions in the modern world. As we saw in the previous chapter, given the right circumstances these skills can still be astonishingly effective. However, there are also occasions when they can be seriously misleading.

One way we learned to survive was by searching for patterns in the world.[2] A flock of birds suddenly taking flight with a cacophony of distress calls, could be perceived as a sign that lions were around. Eating the leaf of a particular plant might be associated with stomach ache. Performing a ritual before hunting might appear to increase the success of the hunt. While many of these apparent patterns might be illusory, disbelieving them usually carried more risk than choosing to believe them. Ignore the distressed birds and you risked being killed by a lion. Getting away quickly when there was no threat simply burned a few more calories. Thus our brains evolved to see patterns everywhere and to believe in these patterns whether they truly existed or not.

In the modern world our need to find patterns, and believe in them, has a cost – we can be seriously misled by randomness. Look closely at the graph below, which shows the monthly demand for a product. After studying the graph, I eventually began to see three underlying cycles in the demand, each lasting about six months – one from months two to seven, the second from months eight to thirteen and the third from about months fourteen to nineteen. Within each of these cycles the demand rises to a peak and then declines. I began thinking about why this might be. Perhaps the product sells well in the summer and is sold in both Europe and Australia, so the six-month peaks coincide with summers in the northern and southern hemispheres. Perhaps the product is an alcoholic drink that sells well in the summer and also around the festive season. If you asked me to use my judgement to forecast the future demand, I'd probably try to extend the pattern I see in the graph by forecasting another six-month cycle that I would expect to end in the twenty-fifth month.

There's just one problem – the apparent 'demand figures' on the graph are actually random numbers generated by a computer. Rather like rolling a 100-sided die, the computer has produced a series of numbers between 1 and 100 where each number is independent of the others. There is no systematic pattern underlying the numbers at all so my forecasts based on the perceived cycle are likely to be well off target.

Not only are we adept at finding patterns where there are none, but we are also brilliant at inventing theories to explain these patterns. Stock market indices often follow random patterns, but watch any business programme on television and you'll find the commentators offer ready explanations for every small twist and turn in a graph – poorer than expected growth figures from China, the resignation of a leading CEO or good weather in California.

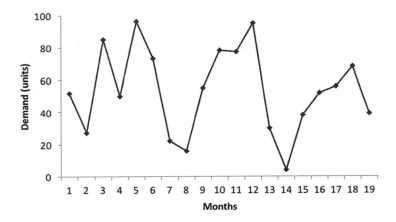

When our theories are challenged we are just as clever at embellishing them. I once sat in a forecasting meeting at a pharmaceutical company. The sales of a drug were slightly up on the previous month but this was probably just one of those random twitches that we see in all sales graphs. 'I think,' said one of the managers, 'our customers are stocking up in anticipation of the forthcoming price increase.' 'But they didn't stock up last time they knew there was a price increase on the way,' retorted the accountant. 'Yes, but interest rates were high then so holding large stocks was expensive. Now they're lower there is an incentive to stock up,' replied the manager. Often, the very last thing we are prepared to do is admit that we don't know why a graph has gone up or down, or to admit to ourselves that it's just the effect of randomness. The Latin phrase *felix, qui potuit rerum cognoscere causas* – happy is he who understands the causes of things – is apt.

Our compulsion to find patterns means that even animals

can outperform us when we encounter randomness. In one experiment, a rat was placed in a maze shaped like the letter T. At intervals, food was placed at the end of either the left 'arm' of the T or the right 'arm'. The sequence between left and right was random, but on 60 per cent of occasions food appeared at the left side. The rat soon learned that the left side was more likely to deliver food and always went to that side so its prediction was correct 60 per cent of the time. When the sequence of left and right placements was shown to Yale University students, their large brains started searching for patterns. Perhaps after two right placements a left placement was more likely, or perhaps after three lefts a right would nearly always follow. Of course, these perceived patterns were false. As a result the students only predicted the correct side 52 per cent of the time. The rats were the more accurate forecasters.[3]

Admitting we don't know what's going on may be anathema, but so is contemplating the possibility we might be wrong. Once we've established a belief that a particular pattern exists, we tend only to recall events which support that belief – a phenomenon known as illusory correlation. If some years ago I bought an unreliable car made by 'The Speedy Company', I might develop an inkling that all cars built by Speedy are unreliable. Every time I see a Speedy car broken down by the side of the road I sigh knowingly, and my belief is reinforced. But I have no recollection of all the other manufacturers' cars that I pass with their bonnets open, or of the hundreds of Speedy cars that zoom down the road taking their happy owners on yet another trouble-free journey. 'Don't buy a Speedy car,' I tell my neighbour, 'I predict you'll have problems.' In fact, if I looked at the data, Speedy might be near the top of the league table for reliability. If we believe that Friday the 13th is unlucky, that the phone tends to ring more often when we're in the shower, that redheads are hot-tempered or that Geminis are socially outgoing, then our experiences will probably only reinforce our beliefs.

Selectively recalling events is a passive process – events simply register with us more often when they conform to the pattern we believe in. But we also have a tendency to actively search for information that confirms our beliefs and downplay any disconfirming evidence. Psychologists refer to this as confirmation bias. Tony Blair and George W. Bush exhibited confirmation bias when they believed that Iraq had weapons of mass destruction (WMDs) in 2003 and decided to invade the country. Huge weight was placed on the evidence of Rafid Ahmed Alwan al-Janabi, an Iraqi defector codenamed 'Curveball', who claimed he had worked as a chemical engineer in a factory manufacturing biological weapons that were a part of Iraq's WMD programme. This was despite earlier misgivings that many intelligence agents had had about Curveball's claims. One suspected he was 'a lying alcoholic ... trying to get his green card ... and playing the system for what it was worth'.[4] The fact that United Nations weapons inspectors could find no WMDs was also discounted, and there was a rush to gather evidence to support the case for an invasion. Any information would do. This even included an American student's PhD thesis that was plagiarised in documents arguing the case for invasion. Even its typographical errors were transcribed and included in what became known as the 'dodgy dossier'. Of course, no WMDs were found.

Confirmation bias has also been found amongst executives who favour a merger between two firms.[5] They actively seek evidence of synergies between the two companies and highlight the forecasts of savings that will result. In contrast, little attention is paid to the costs the merger might incur – such as the costs of getting rid of excess production capacity. Information that provides strong indications the merger will be a failure is also discounted.

Most forecasts don't have the disastrous consequences of the Iraq War or even a failed merger when people get them wrong, but we should still be wary of confirmation bias. For the forecaster who believes they have found reasons for patterns in their data, confirmation bias serves only to intensify these beliefs. And if we've written down explanations for our thoughts then we're even more likely to stand by them, even as evidence that we're wrong mounts up. Michael Cipriano, now based at James Madison University in Virginia, and Thomas Gruca of the University of Iowa observed students who were engaged in prediction markets set up to forecast the success of movies. Those who were asked to write down the rationale for their initial forecasts tended to stick with them even when the latest box office information indicated they should rethink their initial prediction. Students who had not recorded their arguments were less resistant to the new information.[6]

## Stories trump numbers

Our skill at inventing theories to explain what are really random movements in graphs is symptomatic of our close relationship with stories. Stories have been used throughout history to hand down wisdom from one generation to another or to bring colour and meaning to advice about how we should live our lives. We often make sense of the world through stories. One study[7] found that when jurors had to decide on whether a defendant was innocent or guilty, they rehearsed stories of what might have happened to explain why the person was, or wasn't, involved in the crime and how it might have been committed. Our natural inclination towards narrative explanations means that when they clash with indications from numerical data, the story usually wins.

Imagine you are in a company forecasting meeting and the sales graph is showing a 10 per cent increase in sales over the last month. You are told by the forecasting manager, who has a 'first' in Maths, that: 'a computer-based exponential-smoothing forecast indicates that the underlying trend in sales will remain flat, despite last month's high sales. The computer forecast is discounting the recent rise as a random blip.'

The marketing manager jumps in: 'Last month we appointed a dynamic new young sales manager who has an MBA from the Harvard Business School. His energies seem to be endless. These results show that he is already working miracles with our sales. We can now expect a significant upward trend.'

Which account would you go with: the flat trend or the significant upward trend? There's evidence that, to most of us, the dull technical statistical forecast would offer no competition to the colourful inspiring story about the new manager. We would opt for the forecast of an upward trend. And this would be despite the evidence of only one month's sales.

## The perils of imagination and memory

Earlier we saw that we have a tendency to selectively recall events that support our belief that two things are correlated and forget about events that would disconfirm this belief. When we need to judge the probability of future events, such as a terrorist attack, being a victim of robbery or winning a lottery, we are likely to rely again on our memories. A few years ago neighbours of mine won a fortune on the National Lottery. Soon there were shiny, new expensive cars in their driveway. Then fleets of vans arrived to fit their new kitchen, install a hot tub and build an ornamental fishpond. When I recall this, the chances that I'll win one day don't seem quite so slim. In cases like this, the ease with which we can recall similar events occurring in the past acts as our guide to the likelihood that they'll occur in the future. This is not necessarily a bad strategy. More frequent events are more easily recalled and also generally more likely to happen in the future. Most people would correctly give themselves a relatively high probability of suffering from a cold at least once in the next five years. Colds are common; we easily recall instances of ourselves or others having colds and they are likely to continue to be common in the future.

Problems arise when ease of recall is not related to the chances of an event occurring in the future. Some relatively rare events are easily recollected precisely because they were unusual. Or we do so because they were vivid or simply as they had occurred recently. As a result people tend to overestimate the probability of these events occurring again. One source of this bias is the media. Events are newsworthy because they are unusual or dramatic. An ordinary person dying as a result of a fall will not usually make the national news on television, but if they are savaged by a dog and die, the news cameras will soon be heading for the scene of the tragedy. As a result, the probability of being killed by a dog might seem to be higher than the probability of dying as a result of a fall. Yet in the USA in 2009, for example, the probabilities were one in 9 million and one in 15,000, respectively.[8] One study in the USA found that people thought deaths through homicide were more frequent than deaths from diabetes, and that tornados killed more people than asthma.[9] In both cases the second cause of death was more common. Many people fear shark attacks, but in September 2015 the Huffington Post revealed that so far in that year more people had died worldwide trying to take selfies than from shark attacks. Attempts to take selfies had led to people falling down flights of stairs at the Taj Mahal, being gored by a bison in the Yellowstone National Park and being hit by moving vehicles. News about an airline crash might make us fearful of flying. But spending time in our kitchens is much more dangerous.

Terrorists, in particular, exploit our tendency to overestimate the risk of dramatic or media-highlighted events occurring. Their terrible crimes are inevitably headline news and many people live in fear of being a future victim. Yet the probability of being physically harmed is extremely small – roughly 0.00001 per cent of Americans were injured or killed in the 9/11 terror attacks, the worst in US history. Despite this, predictions that they might be victims of terrorism can have a negative effect on people's enjoyment of their lives. While writing this chapter I got talking to an elderly African-American lady at breakfast in a Californian hotel. 'I'd just love to visit Europe – to see Buckingham Palace and all those other historic places,' she said in an accent straight from the Deep South. 'But after 9/11 I thought no way. If they crash the plane in the sea they might never find your body.' In a 1993 study, people were asked how much they were willing to pay for life insurance to cover a flight to London. On average, they were willing to pay $14.12 for insurance that solely covered death resulting from a terrorist attack, but only $12.03 for insurance covering death for any reason.[10]

## Chained to an anchor

We'll see in Chapter 9 that people can be too eager to spot trends that aren't there, but when there really is a trend, are we likely to be better at extrapolating it than a mathematical forecasting method? About forty years ago, the psychologists Amos Tversky and Daniel Kahneman conducted a simple and now famous experiment, involving a group of individuals.[11] The participants were asked to estimate the percentage of African countries that were members of the United Nations. But before they made their estimate, the experimenters spun a wheel of fortune, containing the numbers 0 to 100, to generate a random figure. Let's say the number ten appeared. The participant was then asked whether the required percentage was higher or lower than the generated number. Next, they were asked to make their estimate by moving upwards or downwards from this number. Remarkably, the random number had a significant influence on the estimate. For example, those who saw the number ten generated, typically estimated that the percentage of African countries in the UN was 25 per cent. Those who drew the number sixty-five typically estimated a figure of 45 per cent. Once we have a number in our head it becomes what is known as an 'anchor'. Like the planet Jupiter, anchors have a large gravitational pull which mean it's difficult to get away from them so the participants' estimates stayed close to the random number, even though it was irrelevant to the task. When I told my students about anchoring, some of them thought they'd spotted an opportunity. They wrote '99 per cent' in large characters at the front of their coursework in the hope that I'd anchor on this when deciding their mark.

Anchoring has been found in a huge variety of situations

and it also appears to affect people when they make forecasts by eyeballing graphs that contain trends. It seems that they anchor on the most recent observation in the graph and then adjust from this to try and take into account the upward or downward trend. The problem is that the adjustment is usually insufficient so the effect of the trend tends to be underestimated. Study after study[12] has shown that, when graphs exhibit upward trends, people's forecasts underestimate the rate of growth so the forecasts they make are too low. For downward trends, the problem is even worse and forecasts tend to be far too high.

## A pleasant pension surprise?

Most spectacular is our inability to use our judgement accurately to forecast things that are growing at exponential rates, such as populations, energy use or pollution. Many people have the same problem forecasting the growth in their savings when they are receiving compound interest. As a result, they tend to grossly under-forecast the amount of money they will have in their pension pot when they retire.[13] In this case our inability to handle non-linear relationships may be the source of the bias. We are happier making predictions when numbers appear as a straight line if plotted from left to right on a graph, such as 2, 4, 6, 8, 10, 12 and so on. When numbers grow in a non-linear way, such as 2, 4, 8, 16, 32, 64 …, our predictions are usually much too low.

To illustrate this, assume that you have a sheet of A4 paper that is 0.193 mm (0.0076 inches) thick – dimensions that are fairly standard – and you can fold it forty times. How thick do you think the folded sheet would be? Most people give answers such as one centimetre, or possibly, two. In fact, if we do the calculations we'll get 0.193mm x $2^{40}$ =212,206 kilometres or 131,859 miles. That's more than half the distance to the moon. You might have argued that the sheet was impossible to fold forty times in the first place because it would become too thick after about six folds. The answer shows that it would also be impossible without a ladder into outer space.

In a series of experiments[14] Dutch psychologist William A. Wagenaar and his research colleagues found that intuitive judgement is hopeless when it comes to forecasting phenomena that are growing exponentially. The majority of people produced forecasts that were less than 10 per cent of the actual value. Even people whose jobs involved dealing with exponential growth processes (members of the Joint Conservation Committee of the Senate and House of Representatives of the Commonwealth of Pennsylvania) did not do any better. When it comes to predicting exponential growth: stick with the computer.

## Judging lots of possibilities

Suppose you are asked to estimate the probability that each of this year's short-listed movies will win an Oscar for best picture. For the 2017 award I estimated that *La La Land* had a 65 per cent chance of winning, *Manchester by the Sea* had a

15 per cent chance, *Hacksaw Ridge* a 10 per cent chance and so on. Or imagine you've been asked to estimate the probability that each Olympic finalist in the men's 100 metres will win the gold medal. Typically our judged probabilities in these situations don't add up to 100 per cent. Either they amount to more than 100 per cent, suggesting it's more than certain that one of the outcomes will occur; or they add up to less than 100 per cent, suggesting their forecast hasn't exhausted all the possibilities. In one experiment[15] people were given a record of the previous performances of horses in a race and were then asked to estimate the probabilities that each horse would win. The more horses that were in the race, the more the sum of the probabilities exceeded 100 per cent. This was also true when more information was given about each horse. By focusing separately on each horse's chance of winning it seems that people lost the big picture of the race as a whole.

Another problem comes when people are asked to estimate probabilities for quantities like sales, costs or time. One approach is to ask them for the most likely value and then for a smaller and larger value, so the range (called a prediction interval) has a 90 per cent chance of capturing the actual outcome. For example, I might be asked to forecast how long it will take to build an extension to my house. I estimate that it will most likely take ten weeks. But, if everything goes well, I might complete the work in seven weeks. If I hit snags, the job might take twelve weeks. Overall I reckon there's a 90 per cent chance I'll get the job done in between seven and twelve weeks. All the evidence from research suggests that my range will be too narrow to give it a 90 per cent chance of including the actual building time. If I regularly estimate ranges like this I'm likely to be surprised how often the actual time falls outside them. When Itzhak Ben-David and his co-researchers asked a large sample of US financial executives to produce 80 per cent prediction intervals of one-year-ahead stock market returns, the ranges they estimated should have captured the actual returns 80 per cent of the time.[16] In the event they were so narrow that they only included the true values on 36.3 per cent of occasions.

There are many situations like this where we tend to underestimate the scale of uncertainty we face.[17] One potential cause of overly narrow ranges arises when we estimate the most likely value first. In my building time example this was ten weeks. Once the ten weeks was in my head it became an anchor and my subsequent estimates of the high and low values were likely to be too close to it, leading to a range that was too narrow. But the excessive narrowness might also result from a failure of imagination – I just couldn't contemplate circumstances where the building work would take more than twelve weeks. Or perhaps I just had a problem thinking in terms of probabilities. After all, I took '90 per cent chance' to mean I'd be right 90 per cent of the time – but I'm only building my extension once. So it's likely that I'd have difficulty relating the concept of probability to my one-off project.

In the last chapter we saw that forecasts based on split-second intuitive judgements can be astonishingly accurate in areas where we have plenty of experience and practice. This chapter showed that in other circumstances our judgement can also lead to awful forecasts. If the forecasting problem is unfamiliar, or there are mountains of data to process, or there are complex relationships between what we want to forecast and the factors that influence it and if we need to memorise what happened in the past, then we probably would be better off turning to the computer. We'll look at when you can trust a forecast from a computer – and whether people are willing to believe a computer's predictions – in the next chapter.

# CHAPTER 3

# MORE BYTES THAN
# WE CAN CHEW

**A case for banning ice cream?**

Our intuition can be astonishing when the conditions are right, but as the last chapter showed, our memory, imagination and inability to handle randomness can seriously let us down on other occasions when we use our judgement to make forecasts. In addition, we just don't have the processing power in our heads to handle lots of data and to perform complex mental calculations with this data. In our prehistoric past we would never need to perform calculations in our head like: $55.218 + (2.453 \times 1.843)/(16.128 - 3.456)$ in order to survive. We didn't have to weigh up and combine the messages in a host of economic indicators – some positive and others negative – to assess the probability that the economy would tilt into recession in the next six months. And we were not confronted with the daily prices on the London Stock Exchange for the last 12,000 days and asked if there was a predictable pattern in their movements.

Now we live in an age of big data. Supermarkets and other companies collect detailed micro-information on our purchasing patterns. Data is trawled from our smartphones, fitness trackers and internet activities. Health services store our medical data. Traffic and weather sensors, cameras and radio-frequency identification readers (RFIDs) feed continuous streams of data to remote computers. The world is awash with data. When once we talked about kilobytes of data, specialists now refer to exabytes, zettabytes and yottabytes ($1,000^6$, $1,000^7$ and $1,000^8$ bytes, respectively). The scale of these numbers is difficult to comprehend. For example, it's been said that all words ever spoken by human beings could be stored in approximately five exabytes of data. By 2013 it was estimated that there was 1,200 exabytes of information in the world. To store this on CDs we'd need five separate piles, each stacked to the moon.[1]

One reason for this huge rise in data is a process called 'datafication'. This refers to our ability to convert many aspects of our lives and environment that were previously unmeasured into computerised data. For example, a person's mobile phone activity can be turned into data intended to reflect their personality type. Their social interactions and the extent to which they influence others can be gleaned from their behaviour on social networks and 'datafied'. Text analysis software can be used to transform verbal reports into data. For example, written reports of injuries to police and security guards in Sweden were analysed to find the types and prevalence of the hazards they faced. Even how we brush our teeth, or the way we sit in a car, can now be turned into detailed data.

Inferring reliable predictions from the juggernauts of data now commonly available is far beyond the processing capacity of human judgement. When confronted with such volumes of data we have to resort to examining only tiny sub-samples of it. Even then our inherent biases would be likely to distort our perception of even these small chunks of information. This is the territory where computers can win. Their ultrafast processors can zip through huge datasets uncritically looking for correlations between variables. They are always consistent and they never get tired, bored or emotional. But their lack of knowledge of the world can be both their strength and weakness.

It can be a strength because their blindness can throw up correlations that we would never have expected or even thought of investigating. Data analysed by a San Francisco company suggested that orange used cars are more reliable than used cars in other colours.[2] A US online lender found that people default on their loans more often when they complete their loan application forms using only capital letters.[3] Wal-Mart found that demand for strawberry Pop-Tarts in America increased following severe weather warnings.[4] Other analysts found that people are more willing to answer the phone when it's very humid or, alternatively, when it's snowing or very cold.[5] All these relationships are potentially useful when making predictions.

But there's a danger – just because two things are correlated it doesn't necessarily mean that one is causing the other. There's a strong correlation between Brazil's population in each year since 1945 and the average cost of a train journey in Britain in these years. But that doesn't mean I can blame Brazilians when I get to the station and discover fares have risen again. The two quantities just happened to have increased over the same period for different reasons. Nor would I be justified in campaigning to ban ice cream because a spike in sales correlated with an increase in deaths by drowning. The correlation arises because of a third factor – the weather. When it's hot, people buy more ice cream and more people go swimming. Some proponents of big data have argued that we no longer need to worry about understanding why two things are correlated. For example, if we are running a supermarket and find that, generally, last week's cat food sales correlate with next week's sales of printer cartridges, do we need an explanation for this? Surely, all we need to do is look at last week's cat food sales and we'll get a reliable forecast of how many printer cartridges we'll sell next week. Given our talent as humans for inventing and believing bogus theories to explain correlations, it's argued that worrying about why there is a correlation is a waste of time. 'Forget explanations. Let the numbers speak for themselves' is the cry. 'Petabytes allow us to say "Correlation is enough"', Chris Anderson famously wrote in WIRED magazine.[6]

Certainly, there are reports of successful predictive systems that operate without an understanding of causality. Data from sensors in North Sea oil equipment, which separate sea water from gas and oil, can be used to warn that the equipment is in danger of shutting down. A shutdown costs millions of dollars. It has been found that particular complex patterns of data occur when a shutdown is imminent. The warning system operates without any understanding of why these patterns presage a shutdown. Fire services have been able to successfully predict where fires are most likely to occur, allowing them to target resources for fire prevention to these locations. But it is argued that they don't need to understand why particular features of buildings and their occupants make them more vulnerable to fires.

Richard Berk, professor of criminology and statistics at the Wharton School of the University of Pennsylvania, is an expert in the use of computers to forecast the probability that individuals will commit violent crimes. He argues: 'If the computer finds things I'm unaware of, I don't care what they are just so long as they forecast. I'm not trying to explain.'[7] A Harvard University medical study found that American men aged forty-five to eighty-two who skipped breakfast had a 27 per cent higher risk of suffering from coronary heart disease over a sixteen-year period.[8] Do we need to be able to explain this link before we feel confident in advising men in this age group to rise earlier so that they have time for toast or porridge?

There are parallels here with the differences between invention and science. James Lovelock, a leading independent scientist famous for his Gaia hypothesis, has stated that inventors can seldom provide a rational explanation of how their creations work at the time of the invention.[9] Sometimes it takes years for science to catch up and provide at least a partial explanation. But does this matter as long as the invention works? It does if an invention will only work under certain conditions and our lack of understanding means we are unaware of this. For example, suppose that unknown to its creator or anyone else, a machine is in danger of exploding if it's operated below a certain ambient temperature. In our ignorance we happily decide to use it in freezing weather and disaster ensues.

The same can be true of forecasting. Predicting without understanding might work as long as nothing fundamental changes. But if it does, we can be seriously misled. Suppose a widely reported study finds that people who drink expensive tea tend to live longer. In reality, this simply reflects the fact that only wealthier people, who on average live longer than poor people anyway, can afford higher-priced tea. But the report doesn't say this. After hearing about the report Mary, a person struggling to make ends meet, decides to sacrifice some of her other food purchases to scrape together enough cash to buy a premium brand of tea regularly. Of course, the prediction that this will lead to a longer life doesn't apply to her. Mary's conditions are different. Worse still, the switch from nutritious food to expensive tea might even shorten her lifespan.

Forecasts are made to inform decisions, so decisions based on misinterpreted correlations, like Mary's, can catch us out in other ways. In 1999 a paper was published in the leading scientific journal *Nature* showing that children under the age of two were more likely to be short-sighted if they slept with the lights on at night.[10] The best decision seems obvious – switch off the lights in your child's bedroom. Subsequently, the finding was rejected by other researchers. It was found that parents with myopia had an increased tendency to employ lights as a night-time aid for their children.[11] The children probably inherited their short-sightedness from their parents, not from the lights, though deciding to switch them off would have at least reduced the electricity bill.

## Did Trump thump big data?

As in Britain's 2015 election, which we discussed earlier, the big data analysts were out in force in 2016 trying to predict who would win the US presidential election. With a few exceptions, the clear message they conveyed to the media was: Hillary Clinton will be the 45th US President. And just as in Britain's general election, a shock lay in store for them. Donald Trump hadn't even arrived on the stage to deliver his victory speech before the backlash against predictions based on big data began. 'If "big data" is not that useful for predicting an election then how much should we be relying on it for predicting civil uprisings in countries where we have an interest or predicting future terror attacks?' asked Patrick Tucker, author of *The Naked Future: What Happens in a World That Anticipates Your Every Move*.[12] Britain's general election, Brexit, the

2008 financial crisis, Donald Trump – the high-profile prediction disasters appear to be mounting up. So are the days of big data numbered? Is it a failed technology that should be abandoned, just as the 1937 Hindenburg disaster led to the demise of the era of the airship?

The truth is that, despite its superhuman analytic powers, big data number crunching is not untouched by human hands. It's humans who decide what data to collect to feed their rapacious algorithms. It's humans who design the algorithms, so they embody the prejudices and assumptions – both conscious and unconscious – of their creators. And it's humans who have to interpret and make sense of the mass of charts and figures they disgorge.

I once taught in the mathematics department of a university. Some of my hard-headed mathematical colleagues were suspicious of my research into the role of human judgement in forecasting. 'Surely, anything can and should be predicted mathematically,' argued one, who evidently saw the world divided into two parts: the world of mathematics and the world of fudge. He clearly saw me as a resident of the latter. Yet all forecasting – no matter how sophisticated the mathematics it employs – involves judgement calls. The analysts making election forecasts must judge whether their polling methods are sampling a representative cross section of people who will vote. They have to judge whether people are likely to respond honestly when asked whom they'll vote for. When programming their algorithms they must decide how much weight to give other sources of information such as data from social media, internet searches or newspapers. Or how much notice to take of historic patterns in voting behaviour. Above all, they need to evaluate the mounds of mute metrics and translate them into meaningful predictions of voter behaviour.

Trump's victory caught analysts out because their human judgemental biases betrayed them at every stage. In particular, they paid too much attention to opinion polls, despite these being mere snapshots that attempt to assess voter intentions at a point in time. Plot them on a graph and they can sometimes look as volatile as a heart rate display for a person alternating between sleep and strenuous exercise. Polls can also be problematic, as those whom the pollsters manage to interview aren't necessarily representative of those who actually turn out to vote. For example, it's been claimed that the polls under-sampled non-college-educated whites who tended to support Trump. Even when cornered by a smiling pollster, some people would be too shy to admit they intended to vote for a candidate as controversial as Trump, or that they were implacably opposed to a female President. According to some observers, voters are less guarded in revealing their political views on social media than in polls.[13]

Then there's the data interpretation: if almost everyone else saw a Clinton victory an analyst was likely to view their own data through the same lens. One forecaster admitted: 'Influenced by the percussion of polls and punditry heavily suggestive of a Clinton win, I allowed myself to ignore signs in the data that Trump was ahead in both the battleground states overall and Florida. That was a mistake.'[14] The problem isn't big data; it's the way we humans relate to it.

Any obituaries for the era of big data-based predictions are almost certainly premature. Big data is a relatively new phenomenon. People need to work on improving their use of the technology and, doubtless, they'll learn from these recent disappointments. And in the case of Trump not all forecasts were awry.[15] Even though Nate Silver's model said a Clinton victory was most likely, five days before election day it also indicated there was around a one-in-three chance that Trump would be the winner.[16]

## Overruling the machine

While electoral analysts may be closely wedded to the outputs of their computer models, this isn't the case in many other fields. Sales forecasters and economic forecasters seem only too willing to overrule predictions emanating from their computers. In one study we found that sales forecasters in a well-known food company were changing over 90 per cent of the computer's forecasts – a huge task given the thousands of products they sold.[17] Economic forecasters build huge sophisticated statistical models of the economy to forecast growth, inflation or unemployment. But then they tinker with these models, replacing some of the model's estimates with their judgements. Usually they do this without documenting why they have made the changes.[18]

This may seem odd. In other aspects of our lives many of us are only too reliant on computers and automation. We often unquestioningly accept what a sat nav tells us, sometimes despite the evidence of our own eyes. In 2006 *Reuters* reported that a 53-year-old German driver obeyed his sat nav's command to 'turn right now' – and drove straight into a roadside toilet building causing €2,000 worth of damage. His eyes would have told him the correct turning was thirty metres ahead.[19] In the same year, American band Viscount Oliver's Legendary Four Tops thought they were heading for a sold out concert in Cheltenham, in south-west England. Unfortunately, someone had typed Chelmsford into the device. The band ended up 136 miles away and missed the concert.

Much of this interference with computer forecasts may simply reflect the fact that human forecasters need to justify their pay. If I'm a sales forecaster in a company I might not stay in the job for long if all I do is download computer forecasts before spending my day talking football. So I make lots of small changes to the forecasts. 'The computer says we'll sell 2,080 tubs of margarine, I'll make that 2,079. 4,183 bottles of milk? No way! I reckon we'll sell 4,184.' When a research team I was part of examined thousands of sales forecasts in four large companies we found the vast majority of judgemental adjustments to computer forecasts were small.[20] Because they were small they couldn't do much damage, but they did slightly decrease accuracy. I've even heard reports that, when managers meet together to review the latest sales forecasts from their computers, those forecasts that are discussed earliest have a greater chance of being adjusted. After the managers have changed enough forecasts to justify the meeting, they wave the remaining ones through so they can get back to their offices in time for another coffee.

But even after we've allowed for tinkering it seems that people in some fields are often genuinely sceptical about computerised forecasts. So why do we regard a computer's forecast with more suspicion than the advice of a sat nav? One reason is that a computer's forecast will almost always appear wrong because of noise – random movements in graphs that we can't be expected to predict.

Mr Jones wins a bet and buys a bottle of champagne to celebrate, so sales of champagne go up by one bottle. Mrs Smith, celebrating the birth of her grandson, accidentally drops her champagne bottle in the kitchen and has to return to the shop to buy another. That's two more bottles we've sold, but the computer couldn't have predicted this small increase in sales. However, these apparent errors in the computer's forecasts make us vigilant. We unfairly blame the computer and our faith in its predictions is seriously reduced. This makes us alert to every twist and turn in the graph we think the computer has missed so we are less ready to accept what it's suggesting. In contrast, systems like sat navs produce errors relatively rarely so we begin to perceive they are more or less infallible. According to Nicholas Carr, a writer on technology and culture, in these circumstances we grow lazy. We suffer from what is known as automation bias and unthinkingly follow the sat nav's instructions.[21]

Another reason for disbelieving computer predictions is that, as we'll see in Chapter 9, people tend to discount the value of information from the past. We may think the computer has done a great job in analysing past trends and patterns but tomorrow is a different world. We think we know things that are likely to impact on the future which the computer doesn't. The problem is that, as we access more information about what the future might look like, our confidence in our predictions is inflated far beyond the level which is justified by the value of the information.[22] Even being given irrelevant information increases our confidence while simultaneously making our forecasts less accurate.

Then there's the magical allure of human expertise. Despite being shown that the track record of the computer is superior to judgement, many people will either go along with their own supposed expertise or follow the judgement of someone claiming to be an expert.[23] For example, in an experiment, Australian researchers asked people to use their judgement to make a series of forecasts. After each forecast they were shown what a computer was forecasting, based on a statistical model, and asked to revise their prediction if they thought this would improve it. Despite receiving messages like: 'Please be aware that you are 18.1 per cent LESS ACCURATE than the statistical forecast provided to you', they persisted in staying close to their own original judgemental forecasts.[24]

The future prices of stocks and shares are highly uncertain and pose challenges for both computers and human forecasters. Occasionally a stock market guru will gain a reputation for uncanny accuracy, but this is usually a result of a string of lucky calls, and eventually this luck runs out.[25] Yet, even in this situation our belief in human experts holds firm. With colleagues in Turkey I conducted an experiment where we gave people graphs of Friday closing prices on the Istanbul Stock Exchange and a forecast of what the price would be one week ahead.[26] Half of our participants were told the forecasts were those of a stock market expert (the truth) and the other half that the forecasts had been generated by a computer model. Participants were invited to adjust the forecasts if they thought they could improve them. Everyone received the same forecasts. But those who thought they came from a computer model made significantly bigger adjustments, suggesting they thought these forecasts were less reliable.

## Hindsight hinders foresight

Hindsight can play strange tricks with our memories. It can delude us into thinking we are much better forecasters than we are, so we are much more likely to ignore what a computer is telling us. In 1972 President Richard Nixon was about to embark on a historic visit to the People's Republic of China, which had been a closed country since the revolution there in 1949. There was intense media speculation about what Nixon might achieve during his visit to China and psychologists

Baruch Fischhoff and Ruth Beyth used the opportunity to conduct a now-famous experiment.[27] They asked students to estimate probabilities for events relating to the visit. For example: 'The USA will establish a permanent diplomatic mission in Peking, but not grant diplomatic recognition'.

The students were unaware that after Nixon's visit they would be called back by the researchers to recall their estimated probabilities. The details of the visit and its achievements were by now big news, so the students would have been well aware of what had actually happened. If an event had occurred then most of them believed they had assigned it a higher probability than was the case. The opposite was true when an event had failed to occur. The students were exhibiting hindsight bias or the 'I knew it all along' effect – the tendency to believe our forecasts were more accurate than they actually proved to be. Looking back, events often seem more predictable than they were when we made our forecasts. The 2008 credit crunch now looks as if it was inevitable. I'm sure I had at least a sneaking suspicion it was going to happen. The same is true of Labour's defeat in the 2015 general election and President Trump's victory in 2016.

Hindsight bias doesn't just afflict students put in the unfamiliar position of having to estimate probabilities for political events. In one study,[28] 705 people in the early stages of starting a new business were asked to assess the probability that their start-ups would eventually become operating businesses. Later, 198 of these people who had failed to get an operating business off the ground were asked to recall what their probability estimate had been. Their median estimate was only 50 per cent. But when the researchers checked back they found that the prospective entrepreneurs as a whole had actually made estimates with a median of 80 per cent.

Even people operating in the competitive world of banking suffer from hindsight bias, according to another study.[29] This is a world where success can depend critically on an ability to compare the latest information with one's expectations: 'Why are things different from what I expected? What has changed?' Yet the bankers – even those highly experienced – were not immune. Interestingly, those who tended the most towards hindsight bias received the lowest pay. This indicates that hindsight bias can restrict our ability to learn how to improve our forecasting performance. After all, looking back I can see I'm a great forecaster. What have I got to learn? And why should I pay attention to what a computer algorithm tells me?

## The best of both worlds?

We've seen that our judgements about the future can be brilliantly accurate or hopelessly off target. Computers, on the other hand, can churn out predictions untainted by our cognitive limitations and biases, but they cannot understand the world in the same way as we humans. In 2012, IBM's Watson computer beat the greatest ever winners of *Jeopardy*, a US game show that requires knowledge and the ability to think creatively. A journalist remarked that it had won without 'knowing' it had won.

Back in the 1950s, psychologist Paul Meehl argued that mechanical statistical models were generally more accurate forecasters than human judges. But even he recognised there would be exceptions. He posed the case of a professor who usually goes to the movies on a Tuesday night. A computer might analyse the data on the professor's movements and forecast that there is a 90 per cent chance he'll go to the movies next Tuesday. Usually the computer's forecast is accurate. However, this week we learn the professor has broken his leg and the cast he has to wear is too big for the cinema seat. We correctly predict that he won't go – something the computer couldn't do because it had no relevant data.

So is it possible to have the best of both worlds? Can we harness the power of the computer alongside our knowledge and understanding of the world? When big data is involved we are faced with a dilemma. If the computer suggests the rate of crime in a city district correlates with the sales of margarine there, what do we do? Do we use margarine sales to predict crime hotspots? Or do we dismiss the link as one of those amusing but spurious correlations that sometimes fill the space in newspapers on a slow news day? The danger is that our genius for inventing theories and explanations comes into play. Perhaps margarine sales and crime rates are both linked to poverty. Or more worryingly, perhaps there's something in margarine that turns people into criminals. All we can do is maintain a healthy scepticism. We can monitor the crime predictions based on margarine sales to see if they continue to be accurate. We can hide some data from the computer – such as crime rates in other cities – and see if its predictions hold up there. And we can carry out further investigations to see what might be at the root of the correlation. We don't want to throw out valuable predictors of crime. But at the same time we don't want to be duped into flooding neighbourhoods with police patrols just because people are gorging margarine by the ton.

Nevertheless, there's plenty of evidence that if we approach the task carefully the union of human judgement and computer analysis can prove a winner. One principle is that we should only change a computer's forecast when we have a good reason. This will be the case when we have important new information we know the computer had no access to – the professor's broken leg, for example. Or a scandal affecting a presidential candidate. Or a sales promotion sprung on us by a rival company. If we stick to this policy we are likely to improve on computer forecasts. In the study of company forecasting I was involved in, big adjustments to computer forecasts were generally beneficial. If we make a big adjustment we're taking a risk; get it wrong and the boss might be on to us. So big adjustments are usually only made when there's clear evidence they are needed. Similarly, judgemental adjustments to the forecasts of big macroeconomic models tend, on balance, to lead to better accuracy.[30]

But forecasters must be sparing in their adjustments. Intervention should be the exception rather than the rule. As we've seen, the opposite is usually the case. Sometimes the adjustment is a deceit. Managers have already worked out what they want the forecast to be so they adjust the computer's forecast until it fits this value. In a pharmaceutical company the adjustments were made through the computer system. Managers played around with the parameters of the model – overriding values the computer had estimated to be optimal. Or they changed the length of past data that the model was fitted to. Recent upwards slopes in the sales graph were particularly useful. Fitting the model just to this data made the sales forecasts larger, which would please senior managers. The managers persisted in fiddling with the computer model until they got the forecast they wanted. But they referred to the result as the 'computer-system forecast'. This was convenient when the adjusted forecast turned out to have a large error. They could blame the computer.

Rather than allowing people to adjust forecasts as they see fit, a simple way of drawing on the complementary powers of human judgement and computers is to get both to make independent forecasts and then average the results. If the computer forecasts that an election candidate will win 60 per cent of the vote and a political expert forecasts 80 per cent, then the combined forecast will be 70 per cent. In fact, one of the major findings of forecasting research over the last forty years or so has been that combined forecasts are often more accurate than individual forecasts.[31] The benefits accrue because they allow the forecast to draw on a wider range of information. And, if one forecast tends to be too high and the other too low, the combined forecast will tend to cancel out these biases.

Averaging forecasts of different types has been employed to great effect in recent election forecasts. In the US, researchers involved in a project called PollyVote[32] averaged the forecasts of opinion polls, prediction markets, computer-based statistical models and a panel of experts to predict the share of the vote of the incumbent party's candidate in each of the 2004, 2008 and 2012 presidential elections. The forecasts made just before voting commenced were only 0.3, 0.7 and 0.9 percentage points out, respectively. Alas, you've guessed it, in 2016 that man Trump blotted this excellent record. On the eve of the election, PollyVote predicted Clinton would win 52.5 per cent of the popular two-party vote. She got 51 per cent. But one 'moderately bad' prediction out of many accurate ones doesn't mean we should junk averaging forecasts. And at least PollyVote predicted correctly that Clinton would win the majority of votes. All of this suggests that the computer and human judgement working in tandem – and sticking to roles where each performs best – can often produce more reliable forecasts than either working alone.

Better still, if you can include the judgement of an expert in the mix surely you'll get an even more reliable forecast. After all, isn't that why experts are experts? As the next chapter will show, it's not quite so simple.

# CHAPTER 4

# SPOT THE EXPERT

For its Christmas issue of 1984 *The Economist* magazine decided to conduct an odd experiment. It asked groups of four people to make economic forecasts for the year 1994. What was unusual was the membership of the groups: one consisted of finance ministers, the second comprised chairpersons of multinational companies, while four Oxford University economics students made up the third. But it was the fourth group that drew people's attention – four London dustmen (or garbage collectors as they are called in the USA). The forecasts the groups were asked to make were typical of the sort that are used to inform long-term plans by companies and governments. They included predictions of inflation and growth rates, the price of oil and the exchange rate between the US dollar and the pound sterling.

Ten and a half years later, in 1995, someone at *The Economist* remembered the experiment and decided to check how accurate the forecasts had proved. The result attracted headlines around the world. 'Garbage in, garbage out' was *The Economist*'s own headline. The dustmen had tied with the company bosses as the most accurate forecasters. The finance ministers had limped in last. When it came to predicting the price of oil, the dustmen were the clear winners.

Of course, *The Economist*'s exercise was more a Christmas game than a scientific study. Only sixteen people were involved so the sample size was small and the forecasts related to economic conditions at just one point in time, December 1994. It also turned out that the forecasts of all groups were wide of the mark. For example, on average, the participants predicted that oil would cost $40 per barrel. It was selling at $17 when the forecasts were checked for accuracy. The dustmen and corporate chiefs were just the least-worst groups. And perhaps political – as opposed to economic – forecasting is not a dustman's forte. Clearly thinking it was onto something, in 2011 *The Economist* asked five of them whether Barack Obama would win a second term in office. Eighty per cent of the dustmen said Obama would be kicked out of the White House.

Nevertheless, the headlines generated by *The Economist*'s experiment do chime with the findings of much scientific research. In some areas, lay people can produce more accurate forecasts than experts. And our confidence in many experts' forecasts is simply not justified by their reliability. But what is an expert? How do we recognise one? And when can we trust their forecasts?

## Identifying an expert

Experts often get a bad press and this has long been the case. In 1877 the British Conservative politician and one-time Prime Minister, Lord Salisbury, wrote: 'no lesson seems to be so deeply inculcated by the experience of life as that you should never trust experts'. And famously, Nicholas Murray Butler, who was president of Columbia University for most of the first half of the twentieth century said in an address: 'an expert is one who knows more and more about less and less'.

But true experts have many admirable characteristics. The psychologist James Shanteau, of Kansas State University, has spent his career studying what makes an expert and how they think and perform.[1] He argues that, when forming judgements, experts, unlike lay people or novices, rely on automated mental processes that allow them quickly to recognise patterns. Think of the chess grandmaster simultaneously playing against multiple opponents. He or she strolls past a series of boards, moving a piece on each board apparently without a moment's thought. And they end up winning most of the time. A non-expert would carefully analyse every possible move in turn and assess the consequences. In contrast, the grandmaster effortlessly identifies a pattern and knows intuitively how to respond.

As we'd expect, experts also have a wide-ranging knowledge of their fields, including the latest developments. They are better than novices at discriminating between information that is relevant and that which is extraneous – though even they sometimes fall into the trap of being misled by factors that are immaterial. They can also get to the heart of the matter more quickly than others. This stems from their ability to simplify complex issues. Shanteau quotes a medical special- ist: 'an expert is someone who can make sense out of chaos'. And when a problem occurs that is different from the norm, experts are quicker to spot this. Then, rather than applying the usual tools, they can adapt and use strategies that are tailored to the unusual circumstances.

Of course, when looking for an expert we don't have access to the internal mental workings of people claiming to have expertise. We must rely, in part, on the external demeanour they present to the world. So what in a person's mannerisms might convince us they possess expertise? A key characteristic is the ability to communicate – and to communicate persua- sively. A cynical view is that an expert is someone who has the ability to persuade us they're an expert. I'm a sucker for the smooth-talking bow-tied gurus on late night current affairs programmes. They have a polished argument – embellished with the occasional acronym or jargon word – ready for every challenge. At that time of night I fail to see the questionable assumptions in their assertions or realise they've dodged the interviewer's questions.

Above all, those we see as experts possess a rock-solid self-confidence. Philip Tetlock, a Professor at the University of Pennsylvania, tracked the political and economic forecasts of experts in a now-famous study.[2] He found that they had a vanload of excuses ready in case their forecasts proved to be off target. When they wrongly called the 2000 US presidential election for Al Gore, experts insisted there was nothing wrong with their models. They had simply been fed macroeconomic data that was misleading because it was too positive. This wasn't their fault: the data had belied the fact that people were experiencing a slowing economy so they voted for George W. Bush. Others blamed the 'out of the blue' circumstances of Bill Clinton's affair with Monica Lewinsky which worked against the Democrats. This couldn't have been foreseen, they argued.

Then there was the 'I was nearly right' argument. Experts who wrongly forecast that Quebec would vote to leave Canada in 1995 pointed to the closeness of the result: 50.58 per cent voted 'No' and 49.42 per cent 'Yes'. In other cases there was the 'Just off timing' defence. 'The event that I predicted will occur, albeit a bit later than expected; just be patient'. Or what about the 'I made the right mistake' plea? Typical of this is the argument: 'OK, I forecast too high, but that's much better than forecasting too low'. Experts who predicted the USSR would endure for longer than it did argued that it was much safer to overestimate the Soviet Union's durability than to underestimate it. This is only a selection of the defences categorised by Tetlock. A host of others could be called upon to protect the expert's self-esteem and credibility.

Preserving a veneer of confidence is crucial if experts are to remain credible in our eyes. In 1981 the Humber Bridge

– then the world's largest single-span suspension bridge – was due to open in north-east England. For years there had been controversy as to whether the bridge was needed. Spanning the Humber estuary, it was designed to link the city of Hull to the flat and sparsely populated green acres of north Lincolnshire. In the years before construction commenced other roads had been vastly improved and there were doubts that the traffic flow would justify the huge building costs. In the end the Labour government of Harold Wilson gave the go-ahead after a by-election in a local constituency. It had promised the bridge as an election sweetener to boost its wafer-thin majority in Parliament.

Naturally, in the weeks before the bridge's opening there was much interest in how much it would be used by drivers and whether it would benefit the local economy. Two people whom I knew appeared on a local radio programme to discuss their forecasts. One had done months of careful research. He was honest about the uncertainties in his predictions: 'It was possible that x might happen but y was also a possibility and so was z. This would depend on whether … There wasn't the data available so we couldn't be sure…' It sounded as if he was hedging his bets. He was an academic and clearly unused to engaging with the media. He spoke quietly, punctuating his answers with short silences and caveats as he reflected on the questions. You couldn't help thinking: 'If he's an expert he should know what's going to happen; yet he seems to have little idea.'

Then the interviewer turned to his second guest. What a

contrast. This interviewee confidently unleashed a torrent of figures. 'Well, my research suggests that the mean vehicle flow will be x thousand vehicles per week. This will lead to y per cent annual growth in the local economy bringing z thousand new jobs to the region.' It was impressive. It's what we wanted to hear – optimistic forecasts endorsed with the seal of certainty. Here was a true expert. The programme was worth listening to after all.

But there was a problem. Knowing the second interviewee, I was pretty sure he hadn't done any serious analysis. His forecasts were being made up in the studio. He was a guy who loved being in the spotlight. And he combined this with a perverse delight in 'getting away with it'. He knew little or nothing about traffic flow forecasting or regional economics.

The best way a person can impress us with their apparent expertise is by making an extreme forecast that, amazingly, turns out to be accurate. Suppose you predict that the US economy will grow an astonishing 15 per cent next year, or that we will be in regular contact with aliens on a distant planet within twelve months. Friends take you aside and warn that you may have been working too hard. There are stifled giggles when you enter the coffee bar at work.

And then, incredibly, your forecast proves correct. Suddenly you are a visionary. The phone is hot with calls from the media and your boss hangs on your every word, preparing to invest millions in every scheme you suggest. The people in the coffee bar claim to have believed you all along. Is your new status justified? Not according to Jerker Denrell and Christine Fang of Oxford and New York Universities, respectively. Their research[3] suggests that making an extreme forecast which, against all the odds proves correct, is actually a sign of poor judgement.

How can this be? We tend to envy entrepreneurs who became extraordinarily rich by spotting the future that the rest of us failed to see. However, by definition, extreme events are rare. People who carefully study and analyse all the available information will seldom predict such events, unlike those who rely on hunches. Denrell and Fang argue that forecasters using hunches tend to underweight the underlying rate (the base rate) with which events occur and overreact to any striking, but unreliable, new information that they come across. This leads them to produce more extreme forecasts. For example, it may be that only 10 per cent of movies of a particular genre make a profit at the box office. However an intuitive forecaster, who hears that a few friends have enjoyed the movie, might give it a 95 per cent chance of commercial success and hence pay little attention to the depressing base rate. Of course, these forecasters will occasionally be lucky and appear to have a mystical talent for spotting a winner. But most of the time they'll get it wrong.

Denrell and Fang report the case of the CEO of the Hanmi Finance Corporation. In a *Wall Street Journal* survey, he correctly forecasted high inflation when almost everyone else was predicting a low rate. He arrived at his prediction after visiting a US jeans manufacturer whose production was constantly being outpaced by demand for expensive denims. Such levels of demand, he surmised, meant there was lots of money about so high inflation must be around the corner. However, consistent with Denrell and Fang's reasoning, the CEO's average forecasting accuracy was poor: in the two previous surveys he had ranked only 43rd and 49th out of 55 forecasters.

The problem is that we tend to recall with amazement the rare extreme forecasts that turned out to be true. But we forget the many hunches that proved wide of the mark. This is often reinforced by the media. For example, some media outlets reported that the well-known American psychic and astrologer, Jeane L. Dixon, had predicted the assassination of John F. Kennedy in her *Parade* magazine column in 1956. They paid less attention to Dixon's predictions that World War Three would begin in 1958, that there would be a cure for cancer in 1967 and that there would be peace on Earth in the year 2000. Such was the belief in Dixon's prescience that she allegedly advised President Nixon – who she predicted would serve his country well – and Nancy Reagan. Mathematician John Allen Paulos has given the name the 'Jeane Dixon effect' to our tendency to believe in false prophets, because we recall their rare hits and forget the many misses, According to Denrell and Fang, it means 'we may end up awarding "forecasters of the year" awards to a procession of cranks'.

## Our symbiotic relationship with experts
Our predilection for recalling the successful forecasts of those we regard as experts, while discounting their failures, is

symptomatic of the symbiotic relationship that exists between us and experts. As explained in the Introduction, we feel uncertainty like a pain. People who seem to be experts offer to remove or ease this uncertainty. Their confidence gives us confidence that the future they have predicted will transpire.

Our belief in experts' predictions can endure even when common sense should tell us that the forecasts we're being offered are manifestly useless. In one recent experiment, people were willing to pay for predictions of whether a coin toss would result in heads or tails. They were persuaded after being supplied with a series of envelopes that, by chance, contained correct predictions of their most recent coin tosses.[4] Wharton professor J. Scott Armstrong uses the term 'seer-sucker theory' to explain our willingness to pay for expensive expert forecasts even when there is no evidence the experts are capable of producing accurate predictions.[5] In Armstrong's words: 'no matter how much evidence exists that seers do not exist, suckers will pay for the existence of seers'. His explanation for the phenomenon goes beyond the mere comfort of the certainty afforded by experts. Often, he argues, paying for expert forecasts allows us to avoid responsibility. We are not primarily interested in whether the expert will be accurate or not. Our main aim is to avoid taking the blame if things go wrong. 'I consulted the best expert in the field', we can shout when hauled over the coals for a disastrous decision to launch a new product or a bank-breaking foray into the stock market.

Armstrong cites a study by Joseph Cocozza and Henry

Steadman[6] who analysed predictions by psychiatrists in New York. The psychiatrists were asked by courts to predict whether mental patients posed enough danger to warrant involuntary confinement. Judges in such cases face a dilemma – get the decision wrong and a dangerous patient could be released into the community. Alternatively, a person who is safe might suffer unnecessary confinement. But, although the advice of psychiatrists was sought, there is no evidence they can make such predictions with any accuracy. In her controversial book[7], Margaret Hagen, a professor at Boston University, has argued that the psychiatrists would be just as accurate if they simply flipped pennies or drew cards or put on a blindfold and chose without being able to identify the patients. Yet courts typically base their decisions on these predictions. In the Cocozza and Steadman study they accepted the expert's recommendation on 86.7 per cent of occasions. The researchers wrote that the judges appeared to believe the psychiatrists had a mysterious secret knowledge – magical powers enabling them to predict how patients would behave. Conveniently, this also unburdened them of responsibility if their decision proved to be a serious mistake.

## Hedgehogs and their protective spines

Anyone with a scientific view of the world will find it easy to accept that people claiming mystical powers, like Jeane Dixon, will produce unreliable predictions – hosts of forecasts that are way off target, punctuated by the occasionally lucky strike. But, as in psychiatry, there are many situations where true experts regularly come unstuck when it comes to forecasting. This is despite their training, knowledge and special cognitive skills.

Take stock market forecasting. This is a field where highly knowledgeable people spend their working lives scrutinising charts, economic reports and company accounts in an attempt to predict which stocks we should buy and sell. Yet it seems they'd do just as well if they rolled dice or threw a dart at lists of shares to make their selections. In fact, they might even do better. In two studies, Gustaf Törngren and Henry Montgomery of Stockholm University supplied details of pairs of stocks to finance professionals, such as portfolio managers, analysts and brokers.[8] The experts were asked to predict which stock in each pair would perform best over the next thirty days on the Swedish stock market. For example, would Ericsson's shares outperform Volvo's or vice versa? Random guesswork would, on average, lead to 50 per cent of the predictions being correct. The experts managed an average of only 40 per cent – significantly worse than chance. This study confirmed the findings of many others going back to the 1930s when Alfred Cowles[9] asked: 'Can stock market forecasters forecast?' He found that professional forecasters' choices of shares performed slightly worse than the average share in the markets of that time.

Why does all the sweat and toil of financial experts go to waste and lead to such poor forecasting? The general view is that stock markets are inherently unpredictable. Even if they contain small elements of predictability, as some claim, these are submerged in a frenzy of random price movements that defy the skill of any human being to produce consistent forecasting accuracy. When the world is unpredictable the best an expert can do is to advise us of this unpredictability – to warn of the risks of downsides and the realistic probabilities of gains. But few do. It's in their interests to play down uncertainty and give the impression they have an insight into an inevitable future. In fact, many financial experts seem to live under the delusion that they possess this insight. In the Swedish study (above) and several others, they were shown to be grossly overconfident that their forecasts would be more accurate than they proved to be.

Even where the future can be predicted with some degree of reliability, experts can get it wrong. Experts are human and are as likely as anyone else to suffer from the judgemental biases we saw in Chapter 2. In particular, they often have a particular model of the way the world works and can only view events through this single lens. Economists, for example, often belong to different theoretical camps and are motivated to defend their doctrinal territories fiercely. The world is a complex place and adherence to a single world view can prevent people from stepping back and taking a more holistic perspective of how things interact. In particular, economics has suffered from its assumption that people behave in consistent and rational ways to maximise their self-interest. Fortunately, this mythical, selfish and optimising creature created by economists, known as '*Homo economicus*', has little in common with the *Homo sapiens* we see around town, or in shops, or at work. Real people behave irrationally and make mistakes. They can behave like a mindless herd when panic hits the stock market. They can make impulsive purchases and spend weeks regretting them. They can take silly risks and yet invest huge effort in protecting themselves from tiny threats. But they can also act selflessly and generously when there's apparently nothing in it for them. When an expert's mental model of the world is too narrow for them to see the fuller picture, or when it's based on false assumptions, any prediction that follows is likely to be unreliable.

In economics, experts' mental models are inevitably translated into the computer models they use for macroeconomic forecasting. Early in 2017 the Bank of England's Chief Economist, Andy Haldane, admitted that: 'it's a fair cop to say the profession is to some degree in crisis'. He was reflecting on the forecasts that failed to signal the 2008 financial crisis and those that overestimated the damage to the UK economy in the months that immediately followed the Brexit vote. This was, he said, a 'Michael Fish moment' for economic forecasters. The answer was to move away from 'narrow and fragile' models that assumed people always behaved rationally. What was needed was a broader perspective that embraced insights from other disciplines like psychology.

Philip Tetlock, who we met earlier, has given the name 'hedgehogs' to the sub-species of experts who persist in relying on a narrow view of the world to make their prognostications. He took the term from a famous essay by philosopher Sir Isaiah Berlin, 'The Hedgehog and the Fox'. Berlin's essay was widely read, though he claimed he just intended it as an enjoyable intellectual game. He divides thinkers into two categories: hedgehogs and foxes. Hedgehogs exhibit dogged perseverance in believing in their limited vision of the world. They 'know one big thing and toil devotedly within one tradition'.[10] They bring rigid mechanistic solutions to messy problems. And they believe that history occurs as a result of deep underlying laws that can be abstracted through careful analysis of the past. Predictions can then be based on these laws. This belief in laws makes them susceptible to the use of simple historical analogies. For example, many believed – in 1992 – that the ruling al-Saud family of Saudi Arabia would soon be overthrown. Their analogies were the many revolutions and coups that had removed monarchies in Islamic countries like Iran, Egypt, Yemen, Iraq and Tunisia. But these analogies had limited similarities to Saudi Arabia, where the king controlled huge resources and was protected by a loyal military and police force.

In contrast, cunning foxes know many little things. Their thinking is a blend of assorted theories and traditions. They accept the world is complex and untidy – that it inevitably contains contradictions, uncertainty and things open to different interpretations. In Tetlock's words, foxes are more willing to 'weave together conflicting arguments' before making their forecasts. They also can study issues with more emotional detachment when forecasting. For example, some hedgehogs in Tetlock's study had a hatred of the Soviet Union. This caused them to see Mikhail Gorbachev as a hard-line communist who was using perestroika as a breathing space to allow the USSR to reconstruct itself as a more daunting enemy. Several foxes, with their more detached perspective, were able to assimilate a diversity of views, enabling them to foresee the liberalisation of the late 1980s and the subsequent disintegration of the Soviet Union.

Hedgehogs can seem impressive with their unshakeable belief that they know how the world ticks. Whether libertarians or Marxists, monetarists or Keynesians, atheists or believers they can draw on an in-depth understanding of their doctrines to offer us the certainties we crave. Their arguments are well rehearsed and hence well fortified against dissenting voices. Foxes who offer uncertainty and ambiguity can seem insipid in comparison.

Yet it's the foxes who are more likely to deliver reliable forecasts. In Tetlock's study they trounced the hedgehogs by producing probability forecasts of world events that were much more reliable. And they achieved this without sitting on the fence. They could have played things safe and produced probability forecasts close to 50 per cent but, where appropriate, they were prepared to move their estimates up and down the scale. While the hedgehogs were braver and set their estimates closer to the extremes of 'impossibility' or 'certainty', their lack of caution was often misplaced. This meant they were more likely to estimate high probabilities for events that didn't occur and vice versa.

## Experts against models?

Even foxes aren't perfect. As we saw in Chapter 3, greater reliability can often be achieved if computer-based quantitative models are used instead of expert judgement. But like the Luddites of the nineteenth century who smashed the latest textile machines because they feared the machines would steal their jobs, experts have an incentive to resist the use of models. Not only do such models threaten their livelihoods, in some cases they also threaten their status. This is probably why expert cardiologists in the United States don't use the predictive equations recommended by the American College of Cardiology. They prefer to rely on their own judgement despite the superior accuracy of the equations in predicting the presence of disease and the risk of death.[11] The equations are relatively simple to administer. They require answers to a few straightforward questions about a patient's age, maximal heart rate, whether they have diabetes or angina and so on. A score is then calculated that indicates whether the patient's risk is high, intermediate or low. Two US medical researchers, Eta Berner and Mark Graber,[12] have said: 'We look forward to the day when the information-seeking physician with the "fastest fingers on the keyboard" replaces the "brilliant diagnostician" as a role model for our students.' But if you are that person with the reputation as the brilliant diagnostician you'll probably see the keyboard as a threat.

It's not only expert medical specialists who can be outperformed by statistical forecasting models. I've attended many forecasting review meetings where managers sit around a table to decide whether to accept or change the sales forecasts of a computerised model. Marketing and sales managers have an incentive to demonstrate their expertise at such meetings. Most sales forecasting models used by companies rely on a simple statistical formula. Accepting that the forecasts of such models might be accurate poses a threat to the specialists. So they come up with all sorts of reasons why the statistical forecast must be changed. 'Last week some people were telling me they like our new package design. I think we need to up the forecast.' 'The customer's chief buyer seemed very cagey when we last met. I suspect he's looking for alternative suppliers. I would lower the forecast.' Many of these reasons are based on mere rumour or anecdote. In one pharmaceutical company all the cognitive effort and time invested in the meetings led to changes that improved the accuracy of half the forecasts – but it made the other half worse.[13]

## So when can you trust an expert?

We've seen that it's better to put our trust in a fox's forecast than a hedgehog's. The smart-looking guy confidently advancing a coherent theory to back up his prediction is likely to be less trustworthy than the tentative person who acknowledges uncertainty. Of course we'd need to check that the second person's hesitancy isn't because they don't know what they're talking about. If forced to give a medical prognosis, or a forecast of what a supreme court might decide, I would need to be somewhat vague. But my lack of knowledge would probably be obvious. In some cases I wouldn't even be able to list the possible outcomes. So to trust an expert we need to be sure that they have knowledge, and experience, but also some humility in a difficult world.

It's unfair to judge experts negatively if we force them to give us single number or single event forecasts with no probabilities attached. Asking them to produce single number forecasts of exchange rates ten years ahead, as *The Economist* did, tells us virtually nothing. The experts would be better employed trying to assess the degree of uncertainty and the range of possibilities than a single most likely outcome (more on this in Chapter 8).

The media are particularly culpable in asking experts for neat headline-making forecasts while subsequently spinning new smug headlines when these non-probabilistic forecasts are wrong. I recall when two colleagues were asked by a TV programme to perform an analysis of a scurrilous anonymous letter written by a Member of Parliament denouncing the then Prime Minister, Tony Blair. The TV programme wanted to know who'd written it. They'd heard that my colleagues had developed statistical techniques that could identify individuals' writing styles. By comparing the letter's style to the known writing of a list of suspects it was hoped to unmask the mystery writer. But the intemperate letter was short, so the sample of writing was insufficient to pinpoint one MP for certain. The programme's producers were unimpressed. They wanted a definite name. So, under pressure, my colleagues acquiesced and named a well-known rebel. They were almost certainly wrong. When confronted with the accusation the rebel pointed out that he'd said far worse things about the Prime Minister in public. So why write the letter anonymously?

A far better way of assessing how much trust to put in an expert's forecasts is to examine their accuracy over a large number of similar predictions. Unfortunately, this is not always available. But for one professional group we have voluminous records. And this shows some experts can be excellent forecasters. The group? US weather forecasters. Yet, as a chaotic system, the weather is notoriously difficult to forecast in most parts of the US – unless you live in a desert region. So what makes American weather forecasters so successful? There appear to be several factors that allow them to translate their expertise into outstandingly reliable forecasts. First, they presumably have no motivation to make particular forecasts. Television viewers may groan when they hear there's a 90 per cent chance it will rain on a national holiday, but it's difficult to imagine a weather forecaster overriding their true beliefs because they think one forecast may be more popular than another.

Second, they get lots of practice in producing probability forecasts of how the weather will behave. Third, because their forecasts are usually short-term – a day or so ahead – they get rapid feedback on how well their forecast fared. And they get this feedback repeatedly so they have opportunities to learn. This feedback is also unambiguous. If we forecast how well someone will do their job then we might have mixed signals when, later on, we assess our forecast: John has had lots of creative ideas perhaps but has been poor when dealing with customers. In contrast, when forecasting the weather, there are clear criteria for determining whether or not it has rained in a particular location. Finally, weather forecasters are backed up by massive computing power. This means their forecasts are a blend of hard analysis and experienced judgement – a powerful combination.

When these sorts of conditions apply it's probably safe to trust an expert.

# CHAPTER 5

# GROUP POWER

## Conform or else

The 1980s satirical puppet show *Spitting Image* had a wonderful sketch of Britain's then Prime Minister, Margaret Thatcher, dubbed 'The Iron Lady' by her Russian adversaries. She sits in a restaurant flanked by her Cabinet colleagues as a timid waiter approaches to ask for her order. Inevitably, Mrs Thatcher demands a raw steak.

'And the vegetables?' asks the waiter.

Mrs Thatcher doesn't hesitate. 'Oh, they'll have the same as me.'

Of course, Mrs Thatcher's Cabinet colleagues, made up predominantly of men, were far from being vegetables. Mostly educated at Britain's top schools and universities, they had the intelligence, vigour and guile needed to climb the greasy pole of British politics. But, with their leader, they were about to make some very odd decisions – decisions that suggested an unawareness of the possibilities and risks that lay in the future.

Perhaps the most ill-considered decision was the introduction of the poll tax (or community charge) – a tax designed to pay for local government services such as policing and education. Unlike the tax that preceded it, which was related to the value of a person's house, the poll tax was a flat-rate charge on all adults and therefore took no account of their earnings or wealth. Also, for the first time, young adults living at home with their parents were required to contribute to the cost of local government services. The Thatcher government was blindsided by the dangers of the new tax, failing to see how unpopular and unworkable it would be.

Days before it was due to be introduced, a demonstration against it turned into one of the worst riots witnessed in modern Britain. Cars were overturned, shops looted and buildings set on fire. Over a hundred people were injured and more than three hundred arrested. When the tax demands finally landed on people's doormats, huge numbers refused to pay and one police force announced it would be physically impossible to take action against so many defaulters. But Mrs Thatcher persevered, her determination buoyed by past victories over striking miners and Argentinian invaders of the Falklands. This time events would not favour her. The Iron Lady's popularity plummeted and her colleagues – the 'vegetables' – began to see her as a liability. Her refusal to back down was seen as evidence that she was out of touch with voters. In the end, the debacle contributed significantly to the demise of her premiership in 1990. Her successor, John Major, rushed to announce the abolition of the poll tax in one of his first speeches as Prime Minister.

So how did the collective intelligence of top-ranking politicians, supported by elite civil servants, lead to such a disastrous inability to foresee what might transpire? We might expect a group of people to make better forecasts than individuals. After all, a group will bring to the table a variety of different perspectives, diverse expertise and a wider range of information than an individual. A group also affords the opportunity to exchange ideas and challenge and criticise each other's arguments. Only good arguments should survive such attempts to test them to destruction. The forecasts of what might happen if particular strategies are followed should therefore be well founded.

That's the theory, but the practice is often very different. The problem is that members of groups often feel a pressure to conform to what they think the rest of the group is thinking. This was vividly demonstrated in the 1950s in a series of experiments by psychologist Solomon Asch.[1] Let's imagine you've been invited to participate in a replication of one of Asch's experiments, which you have been told is about visual perception. Feeling a little apprehensive, you show up at the appointed time to find a number of other participants sitting around a table, shuffling nervously and apparently wondering what the task will involve. You find a seat at one end of the table. Then the experimenter appears and hands each person two cards. It all looks simple. The card on the left has three vertical lines of different lengths labelled A, B and C. The card on the right has a single line which appears to be exactly the same length as line C. You are told that everyone has the same pair of cards.

'Now,' says the experimenter, 'I'm going to go around the table asking you each in turn, which of the lines A, B or C is the same length as the line on the other card?'

'My goodness this is easy,' you think. 'The answer is obviously "C". It's amazing what these academics can get away with. I expect this guy has got a lucrative grant for this work.' For a moment you vaguely recall reading in the newspapers about some academics who received a large grant to discover why people visit coffee shops. Their conclusion, after extensive research: they go there to drink coffee.

But then the first person gives their answer – a confident 'A'. You wonder if this person's spectacles need replacing. Then the second person says 'A' as do the third, fourth, fifth and sixth. Now it's your turn. What do you say? It seemed obvious that the answer is 'C', but perhaps there is something wrong with your eyesight. Surely all the others can't be wrong. If you say 'C' you risk making a fool of yourself.

Typically 32 per cent of people in your position agreed that 'A' was the correct answer, despite the evidence of their own eyes. What they didn't know was that all the other participants were in cahoots with the experimenter – told to pretend they were participants and to give the same wrong answer. The experiment had nothing to do with visual perception. It was a test to see if people conform to the prevailing view of a group, even if this view is manifestly wrong. And most did – repetitions of the experiment showed that 75 per cent of people conformed on at least one occasion.

When asked later why they had given an obviously wrong answer, some people said they realised they were wrong but did not want to appear 'peculiar' in front of other group members or to be ridiculed. Others said they thought the answer being given by the fake participants must be correct.

## Groupthink is at large

There's a big difference between people judging the lengths of lines in a psychological experiment in the 1950s – when Western societies had yet to become more individualistic – and senior politicians in the late 1980s assessing the likely success of a new tax. But there are good reasons why the politicians might still have a tendency to conform. Concerned by a number of poor decisions by US governments, such as the failure to foresee the Japanese attack at Pearl Harbor in 1941 and the disastrous Bay of Pigs Invasion of Cuba in 1961, psychologist Irving Janis set himself the task of investigating what had gone wrong. This resulted in his theory of 'groupthink', which he published in *Victims of Groupthink*.[2] Groupthink occurs when there is an excessive tendency for members of a group to agree with one another in order to achieve a consensus. It is particularly endemic in insulated groups who have been working together for some time and have become a cohesive team. And it's exacerbated when the group is under stress or feels threatened and when it has a directive leader – just like Mrs Thatcher.

It's easy to imagine what a meeting would feel like when groupthink is casting its shadow. The group is under pressure to reach a decision and its domineering leader clearly favours a particular course of action. Common sense tells you that what the leader is proposing is highly risky. But you've been a member of the group for several years and you all get on well so you wouldn't want to rock the boat or create any bad feeling. Everyone else seems to be agreeing with the leader's preferred choice and rationalising why it should be implemented. Perhaps your doubts are unfounded – it's probably best to keep them to yourself. As the unopposed arguments in favour of the leader's choice pile up, the group begins to think that it's invulnerable – there are no risks. Nor is there any need to seek outside opinion or gather more information before reaching the decision. The group is confident that what it's doing is right and moral. Why bother considering alternative courses of action? A glow of excessive optimism about the future pervades its thinking.

Many of these characteristics would be recognised by anyone observing the government team that met to plan and recommend the poll tax.[3] Chaired by aristocratic old Etonian William Waldegrave, it met in isolation, had no interest in speaking to sceptical outsiders, and kept the financial experts in the Treasury out of the loop as much as possible. Alternatives to the favoured tax were never seriously appraised. The final decision to implement it was taken during a meeting at Chequers, the Prime Minister's country residence. Despite the huge implications of this radical change in local taxation, no papers were circulated in advance of the meeting. Nobody asked questions or raised doubts. After all, everything would be fine.

Of course, disastrous decisions fuelled by groupthink's excessive optimism did not end with Margaret Thatcher's government. More than forty years after Irving Janis published his book, top managers in major corporations are falling into the same trap of grossly underestimating future risks and opting for astonishingly rash courses of action. This may account for the shocking revelation in 2015 that Volkswagen had been putting software into its diesel engines to fool testers into thinking a vehicle's emission levels met regulations.[4] The software reduced emissions when it detected that testing was being carried out. But it would then restore emissions to normal levels – up to forty times above the legal limit – when the vehicle was driven on the road, thus producing more power and greater fuel economy. The collapse of Swissair[5] in 2002, then one of the world's ten largest airlines, as well as the struggles of British Airways and Marks & Spencer in the 1990s, have all been attributed to groupthink. The last two companies had been considered 'darlings of the stock exchange' before they both embarked on highly risky and ruinous globalisation strategies. According to Jack Eaton's analysis,[6] their history of success encouraged managers to think they were invulnerable.

In Marks & Spencer's case, sceptical opinions in the City and the financial press about the company's strategy were ignored. So-called 'facts' on which senior managers based their decisions went unverified. Dissenting voices were discouraged – anyone challenging the prevailing view was seen as a fool. Despite a significant fall in the company's share price in 1995, its then chairman, Sir Richard Greenbury, flaunted Marks & Spencer's invincibility: 'We can do anything we like. It might upset the shareholders but if we really wanted to savage the marketplace we could push up sales by 25 per cent and let profits slide by £800 million.'

Excessive optimism indeed, as by 1998, sales, profits and its share price were nosediving. Actions such as the expensive takeover of Brooks Brothers' stores in the US – many of which hadn't seen a lick of paint for thirty years – were starting to have adverse effects. Back in the UK, competitors like Next had recognised that success in the clothing market lay with younger customers buying less durable garments, rather than the traditional staid products that Marks & Spencer's sanguine managers were offering.

In the years since Janis produced his theory of groupthink, researchers have developed a richer understanding of why groups can make poor forecasts and decisions. In face-to-face meetings, groups often converge on views that are expressed early on regardless of their merits. Asch's experiments show we are prone to defer to the supposed knowledge of others and don't want to appear difficult or foolish by disagreeing.

Also ideas that are expressed early, and in vivid terms, come to dominate our thinking. As a result, they hinder our ability to recall other relevant information.

Even when people can recall information that they hold privately, they may be reluctant to share it. There's evidence that those attending meetings prefer to hear people talk about information that is already common knowledge to most of them. And they give this information greater credibility. As a result the group fails to draw on the potentially crucial information held by silent individuals.[7]

Extreme forecasts are another potential outcome of group discussions. Suppose that a group is to meet and agree on an estimate of the probability that a military coup will take place in a particular Asian country within the next year. If we polled the group members privately before the meeting, we might find they have an average estimate of 35 per cent – so the general view is that a coup is relatively unlikely. As a result the initial discussion in the meeting is likely to be skewed towards reasons why there won't be a coup. This reinforces people's scepticism about the coup's likelihood. Combine this with group members' reluctance to dissent from the prevailing view and by the end of the meeting its participants will be confident in agreeing that the chances of a coup are only 10 per cent. Hence the meeting has served to move people to an extreme position. If, instead, group members had started with an average probability of 65 per cent, it's likely that would have moved to 90 per cent by the end of the meeting.

Of course, most cases of malfunctioning groups go un-documented, but they have probably been the source of unjustified optimism and reckless risk-taking thousands of times in different decision-making situations and organisa-tions.[8] Good forecasts illuminate the wide range of outcomes that might show up in the future, so we can take these con-tingencies into account when making decisions. Groupthink, and other characteristics of face-to-face groups, encourage the opposite. They blind us to most of the possible outcomes and restrict our focus to the one we most want to see. Un-certainty is pooh-poohed. This is a pity, because groups have so much potential for improving forecasts. In fact, as we'll see, given the right conditions, forecasts from groups can be remarkably accurate.

## Ice-breaking

The city of Nenana lies at the heart of Alaska's dramatic land-scape of mountains and white-water rivers and is about halfway between the Pacific and Arctic Oceans. Although pleasantly warm in the short northern summer, its population of about 350 people can see temperatures plummet to minus 30 degrees Celsius on dark winter days, when the sun only rises for about four hours. During the early 1900s, people would eagerly await for the ice to break on the frozen Tanana River. This signalled that riverboat travel could resume and warmer months were on their way. Gambling was a major form of entertainment in Alaska at the time and in 1905, five people placed bets on the date and time when the ice would break. They had unwittingly started one of the most durable annual forecasting competitions in the world.

Now known as the Nenana Ice Classic, the competition has been running in its current form since 1917 and attracts around 200,000 bets each year. The jackpot in 2014 was a record $363,627. All proceeds go to volunteer and non-profit organisations. Each year a striped wooden structure called a 'tripod', even though it actually has four legs, is placed on the ice and connected by a pulley system to a clock on the river bank. When the ice melts in spring, the tripod falls into the water and the clock records the date and time to the nearest minute. In the years of the competition this has always happened between 20 April and 20 May.[9] Bets on the exact date and time when the clock will stop must be placed between 1 February and 5 April. The person whose forecast is closest to the actual time takes the jackpot. Unsurprisingly, given the sums involved, the tripod and clock are continuously monitored and illuminated, just in case someone decides to stop the clock at a time that suits them.

Just forecasting the date when the ice will break on the Tanana is challenging, let alone forecasting the exact time. There can be huge swings from year to year. In 1992 it was as late as 14 May; the following year it was as early as 23 April. But underlying these wild swings is a trend. Over the last few decades Alaska has seen an increase in average winter temperatures double that in the rest of the world. This has meant the date when the ice breaks has typically been getting slightly earlier – by an average of one day every eight years.

The Nenana Ice Classic thus provides an opportunity for a natural experiment to test whether a large group of people can make accurate forecasts of a volatile event when their individual judgements are aggregated. Karsten Hueffer of the University of Alaska at Fairbanks and three fellow researchers decided to investigate.[10] They gathered the forecasts of large subsamples of competitors for each of the fifty-two years for which records were available between 1955 and 2009 (records for the three years, 1963-1965 were not available). For each year they obtained a group forecast by calculating the median of the dates that had been predicted. They then compared this with the forecasts of a range of statistical methods. The methods ranged from simple moving averages, which tried to track the trend, to more sophisticated computer-based models that took into account past relationships between the date the ice would break and weather conditions during the period from January to March. None of these statistical methods could beat the aggregate prediction of the competitors – the so-called wisdom of the crowd.

Further analysis revealed that the group's forecast reflected both the long-term trend and local weather conditions in the months preceding the ice breaking. This is remarkable – individually we would expect the competitors to have difficulty in recalling past data on such variables as temperature, snowfall and snow depth. We would expect them to face even greater problems in combining and processing this data in their heads, in order to convert it into a forecast. But it seems that the crowd, as a whole, is able to make good use of all the available information – and to use it more effectively than computer algorithms.

## Independent minds

Why can groups perform so well in some circumstances, and yet be so alarmingly dysfunctional when it comes to cases like the poll tax decision? The Nenana Ice Classic has a number of features that are known to be conducive to reliable forecasting. The competitors are a diverse group located across Alaska and beyond; in fact people from all over the world now participate. Collectively this means they will have access to a multitude of information sources and will also bring different skills and knowledge to the crowd's forecast. The prospect of a $300,000 prize is also likely to motivate them.

Significantly, each person's forecast is made independently, without any knowledge of what the thousands of other competitors are predicting. This means that conformity, the big enemy of group forecasting, is absent. There are other well-known cases where independence has led to highly accurate predictions by groups of people. James Surowiecki's book, *The Wisdom of Crowds*,[11] recalls the visit of Sir Francis Galton to an English village fair in 1906 where an ox was on display. People were being invited to take part in a competition to predict the weight of the ox after it had been slaughtered. Galton, a half-cousin of Charles Darwin, and a major contributor to the development of statistical methods, analysed the 800 entries and found that none of them predicted the correct weight of 1,198 pounds exactly. But, when he calculated the average prediction, he was surprised to find it was only one pound out.[12]

Why is independence so important? It is likely to lead to a wide scattering of forecasts, but the average of these forecasts tends to be closer to the true value than most of the individual forecasts. In the Nenana Ice Classic there will be both optimists who predict a date that is far too early, and pessimists who think the ice will take an age before it shifts and stops the clock. When we average their predictions their errors tend to cancel each other out. Imagine instead if one person publicised their forecast and it was then mistakenly believed they had some mystic knowledge of when the ice would break. If everyone conformed to this forecast we would lose the power of the group, as we would only have one prediction. If this forecast is wrong there would be no counteracting forecasts to reduce its effect. Even when we don't have this extreme conformity, the power of groups is reduced when people's forecasts are similar.

We'll see in Chapter 7 that some economic forecasters tend to herd together to protect themselves from the risk of looking foolish if they deviate from the views of the crowd. But there are other reasons why people often follow each other's predictions. In companies, sales staff are likely to have access to the same information and to share their views in the coffee bar or at meetings. They may also have similar levels and types of expertise. As a result their sales forecasts are likely to cluster around a particular value. Even 'mild' influence between group members can be detrimental to accuracy. In 2011, as part of an experiment[13] at the Swiss Federal Institute of Technology (ETH) in Zurich, participants had to estimate quantities such as the length of the border between Switzerland and Italy and the number of murders in Switzerland in 2006. Initially they made their estimates independently and the averages of these were quite accurate. But then they were given information about the estimates of other group members and invited to reconsider the value they had originally put forward in the light of these estimations. Unsurprisingly, the estimates of different group members converged – they became less diverse. The problem was that they often coincided around anomalies that were a long way from the true value. Yet, despite this reduction in accuracy, group members became more confident about their estimates after revising them.

The Nenana Ice Classic group predictions were based on the average of thousands of competitors' forecasts, but when individuals' judgements lack independence there is little point in obtaining forecasts from large numbers of people. Each additional forecast will add little new information to the pot, so very small groups will often suffice. One study[14] looked at forecasts made by thirteen *TIME* magazine personnel, regarding the number of pages of advertising that the publication would sell in a year. Researchers examined the effect on accuracy of basing the group forecast on an average of two people's predictions, then three, then four and so on. The improvements in accuracy petered out when the number in the group exceeded five.

## Can we improve on independence?

What we've seen so far suggests that groups composed of independent forecasters are the bee's knees. But sometimes we can do even better. Recent research indicates that if we want to avoid the risk of a group producing a forecast with a large error we should go to the opposite end of the spectrum from conformity and include people who will actively disagree with others. In other words, rather than being independent, the forecasts of groups are improved if some members deliberately make predictions that are negatively correlated with those of their colleagues.[15] People regarded as mavericks can play a useful role here. But even throwing the forecast of a layperson into the pot with those of the experts might improve the group's accuracy, if the layperson diverges from the prevailing view. This is yet another example of where ignorance can be beneficial in forecasting.

How can this be? Clintin Davis-Stober of the University of Missouri and his co-researchers use the analogy of holding a financial portfolio. Here it's often advisable to hedge our bets by including some assets that perform well at times when others are performing poorly – shares in a rainwear manufacturer and those in a sunscreen producer are a frequently quoted example. The two counterbalance one another and lower the risk of the portfolio. If we only invested in the rainwear manufacturer we would experience boom years and years of huge losses, depending on the weather. If we invested in both companies our returns would be less volatile, as losses incurred by one would be compensated by profits in the other. The mathematical model developed by Clintin Davis-Stober and his colleagues assumes that big forecasting errors are much more serious than smaller ones, so we want to avoid any risk of the group's forecast missing the actual outcome by a mile. Hence the counterbalancing of group members' forecasts will have a moderating effect. The average of their forecasts is unlikely to be an extreme number, so there will be less danger of a disastrous forecast error. Note, however, the big assumption that occasional large errors are something we are desperate to avoid. In some circumstances we may be able to tolerate these as long as the bulk of our forecasts are accurate.

There is another problem with independence. If we metaphorically lock each member of the group into a private voting booth so they have no contact with other members then we will avoid conformity for sure. However, we will also deny them the opportunity to share information and to debate and test their arguments against those of others. When people come together they can potentially learn from each other's knowledge and wisdom. Forecasts based on false premises can be exposed through debate. A wider range of possible outcomes can be envisaged based on the different perspectives of participants.

So we have a dilemma. Hold a face-to-face meeting to make the forecasts and we might be lucky and gain the advantages of information exchange through debate. But we also risk conformity and domination by some individuals who are more vocal or senior. Worse still, company forecasting meetings that I have witnessed have sometimes degenerated into mutual recrimination sessions, as people blame each other for disappointing sales. The purpose of the meeting – forecasting – gets lost. For all these reasons Wharton School of Business professor Scott Armstrong has advocated the elimination of face-to-face meetings from the forecasting process.[16] He even quotes the Pulitzer prize-winning humorist Dave Barry: 'If you had to identify, in one word, the reason that the human race has not achieved, and never will achieve, its full potential, that word would be meetings.'

So can we have our cake and eat it? Can we avoid conformity while also allowing interaction amongst group members? Three methods appear to work well.

## Supermeetings

In 1963 American psychologist Norman Maier wrote a book that drew on existing research to propose a set of guidelines for running problem-solving face-to-face meetings.[17] These included allowing time to explore a range of possible solutions, encouraging dissent and protecting individuals from personal attacks when their viewpoint diverged from that of the majority. More recently, Cass R. Sunstein of Harvard

Law School and Reid Hastie of the University of Chicago have made additional suggestions.[18] They argue that the leader of the group and other senior members should encourage candid responses and then remain silent – lest their status distorts the views of other members. Devil's advocates and contrarian teams that deliberately argue for views which oppose a group's position can also be a valuable counter to conformity.

Armstrong claims that approaches like these are rarely followed. But there is recent evidence that, when they are, they can lead to highly accurate forecasts, particularly when a meeting consists of a diverse group of open-minded, reflective, inquisitive and numerate individuals who share a common purpose. People who are prepared to agonise over whether a probability forecast should be set at 35 per cent or 36 per cent – people who are referred to as 'superforecasters'.

In their book, *Superforecasting*, Philip Tetlock and Dan Gardner report how groups of such individuals were asked to produce probability forecasts for world events. For example, members were each asked to estimate the probability that Xiomara Castro would win the 2013 presidential election in Honduras (she ultimately lost to a conservative rival, Juan Orlando Hernández). Often the groups worked closely together to systematically research the background to the event and to gather information from a wide range of sources. When they came together (often online, rather than face to face) people were encouraged to challenge each other in a respectful and non-confrontational way. And they were able to admit, without embarrassment, when they had no idea what would happen or were unaware of key facts. The accuracy of their group-based probability forecasts was impressive and they were significantly more reliable than a highly regarded competing method known as prediction markets, which we will discuss later.

The teams of superforecasters may have been brilliant, but replicating their success is likely to be difficult in many organisations, as Tetlock and Gardner acknowledge. People with the attributes of superforecasters are rare. And it's doubtful how many of us could convert to being such open-minded, inquisitive and numerate individuals through training or other methods, though the notes on the cover of *Superforecasting* are more sanguine. If, say only 2 per cent of us have the talents of superforecasters, we'll need a large pool of people before we can find an elite team – and those left out may be resentful.

But there is some recent evidence from a large study by Pavel Atanasov of the University of Pennsylvania and seven of his research colleagues, suggesting that even lesser mortals working together can produce reliable forecasts if their interactions are well organised.[19] The researchers argue that this can be achieved if we follow five steps. Firstly, allow people to freely exchange information and arguments, but then ask them to submit their forecasts individually for subsequent averaging. Secondly, allow them to revise their forecasts if they want to, after hearing new arguments or seeing the latest information. Thirdly, if we have a track record of people's past accuracy, the group average will be likely to yield a more reliable forecast especially if it gives greater weight to more accurate forecasters. Fourthly, people's track record should also be fed back to them to help and encourage them to improve over time. Finally, add a magic tweak to the group average. When estimating probabilities for future events, the averages of the superforecasters tended to be too close to values suggesting a sitting-on-the-fence attitude – unlike the more disorganised groups we met earlier, who tended to move towards extremes. For example, where there are two possible outcomes – a Democrat will win the next US presidential election or they will not – the superforecasters' group average would tend to be closer to 50 per cent more often than it should be. The Atanasov research team applied a tweak factor, estimated from previous performance, to their groups' averages to push them towards either 0 per cent or 100 per cent. This helped to overcome this bias and further improved the teams' forecasts.

In both *Superforecasting* and the Atanasov studies, members of groups were initially strangers involving themselves in a research project for personal satisfaction or a small payment. Their jobs weren't on the line and they didn't have their boss sitting next to them in their meetings. They also had diverse backgrounds, and so hadn't spent the last ten years working in the same organisation immersed in the same culture and beliefs. In many meetings these advantages won't apply, so the danger of conformity remains. In this case there are two further approaches that may help us to tackle the independence-interaction dilemma.

## Delphic wisdom

In the 1940s, as the world shifted from the horrors of World War Two towards the tensions of the Cold War, US General Henry 'Hap' Arnold saw the need to forecast future technological developments so that they might be harnessed by the military. The war had seen massive advances in technology. Jet fighters, rockets and nuclear weapons had rendered much of the pre-war arsenal of military hardware obsolete, and early computers were harnessing intelligence from seemingly impregnable enemy codes. In 1946, Arnold persuaded the Douglas Aircraft Company to develop Project RAND (an acronym for research and development) to troubleshoot the problem. It was soon found that mathematical methods, such as extrapolating trends, were unsuitable for the task – new inventions do not follow smooth mathematical curves over time.

A method that could draw on the views of different groups of technical experts in the most efficient way was needed. By the 1950s, Project RAND had become the RAND Corporation, a non-profit think tank, and it was there that futurologist Olaf Helmer, philosopher Nicholas Rescher and the mathematician and philosopher Norman C. Dalkey developed the now-famous Delphi method. It was named after the venerated oracle in ancient Greece where an entranced priestess was believed to deliver messages and prophecies from the god Apollo. Apparently Dalkey and his colleagues had the name forced on them, much to their displeasure. To them it suggested that their method was something to do with the occult.

The idea behind the Delphi method is simple. It guards against pressures to conform by asking between five and twenty participants,[20] or panellists as they are known, to make initial forecasts secretly and anonymously. For example, experts in space travel may be asked to forecast the year when a human will first land on Mars. Or the panellists may be asked for a probability forecast, along the lines of: what is the probability the Democrats will win the next US presidential election? Ideally, responses should be accompanied by anonymised arguments supporting the forecast.

Each panellist then submits their forecast and arguments to a facilitator who collates and analyses them. Statistics summarising the panellists' views, such as their median forecast, are fed back to them, together with notes encapsulating their supporting arguments. The panellists, still acting anonymously, are then invited to revise their forecasts in the light of the feedback, but only if they think this will improve their accuracy. The process continues until a consensus is reached or no one chooses to revise their forecasts. Usually, this takes two to three rounds. The median of the panel's responses in the final round is then used as the group's forecast.

While the anonymity of the forecasters is designed to prevent conformity, it can be seen that Delphi still allows some interaction amongst forecasters. When the feedback is circulated they can see what the group as a whole is thinking and study other people's arguments. This doesn't quite have the dynamism of a supermeeting. By banishing direct contact between members we lose some of the advantages of face-to-face meetings. In well-run meetings, arguments can be instantly critiqued and debated. And new information can be sought and introduced at exactly the right moments as the joint search for the appropriate forecast progresses.

But there are still many advantages to using the Delphi method. Panellists can change their minds between rounds without embarrassment or loss of face. In a meeting, as I listen to other people's arguments, I might begin to realise my initial views were wrong. But I might not admit it. If I were to do this, I could risk looking weak or ill-informed. Instead, I'll try to tough it out – desperately searching for bogus arguments or biased information to preserve my position. In Delphi, who cares if I change my mind? No one knows who I am, so I may as well be honest.

Research suggests that the Delphi method is a reliable way of obtaining accurate forecasts from groups.[21] But it must be applied appropriately. Many applications simply feed back the statistics of the panellists' current forecasts. They omit the underpinning arguments. Without these arguments why should I, as a panellist, change my mind between rounds, just because my forecast differs from the group's median? I may do so because I think they must be better informed than me, but there is nothing in the feedback to tell me this. And confidence in one's forecast is often uncorrelated with how accurate that forecast is.[22] Without access to arguments, people are likely to behave as they did in the study by the Swiss Federal Institute of Technology that we discussed earlier. Recall that, when they only had access to other people's estimates of things like the length of the Swiss-Italian border, people simply converged – they instinctively followed each other and their final estimates ended up clustering around inaccurate and often extreme values.

## Market forces

A thing that Delphi panels often lack is incentives – incentives to take part in the first place and incentives to think hard about one's forecasts. Because the Delphi process takes place over several rounds, experts might resent being asked to reconsider their forecast yet again, when the latest feedback arrives. There's a danger that many will decide to drop out at this stage, so we end up having to base the final forecasts on a depleted group of stalwarts. Even if people stay with the process, the incentive to put serious thought into their individual forecasts may be absent. Anonymity can work both ways. If no one knows who I am I'll get no recognition even if I produce an amazingly accurate prediction. And the accuracy of such a prediction would probably be diluted anyway, when it's combined with those of other panellists. On the other hand, responsibility for a potentially disastrous forecast is shared. If I have nothing personally to lose from the forecast being wildly inaccurate, why should I, as an anonymous member of a panel, worry about the risk of catastrophic errors?

One group forecasting method that does offer direct rewards for accuracy is the prediction market. In these markets, assets are traded like stocks and shares. Typically an asset will pay out a specified sum, say $1, if a given event occurs by a specified date. If the event fails to materialise in time, then the asset is worthless.

For example, in November 2008, one well-known prediction market the Iowa Exchange Market (IEM), operated by the University of Iowa, invited people to make forecasts of the box office ticket sales for the movie *Twilight* – an American vampire romance film. The IEM offered the following asset: it would pay out $1 if the official box office ticket sales from 18 November, when the film was first due to be screened, until 21 November 2008, were greater than $90 million and less than or equal to $100 million. If ticket sales did not fall in this range then the asset would hold no value. (Other assets were also offered, allowing people to speculate that sales would be captured by different ranges.) At midnight on 12 November the price of the asset was $0.20. This price indicates the market's view at this specific time, suggesting that there was a 20 per cent probability the movie's ticket sales would fall within the specified range.

Suppose we assumed that the true probability was higher than this – say 40 per cent. It would have been worth buying the asset because we would reckon that our chances of earning $1 would be 40 per cent, providing us with an expected payoff of $0.40. And we could buy the asset for the bargain price of $0.20. If we and several other people decided to buy several units of the asset, its price would rise. It would keep on rising until the market decided that no potential profit was to be made by further purchases. If everyone thinks the same as us, this will be when the price hits $0.40, indicating that the market now believes the true probability is 40 per cent.

In the event, the price of the *Twilight* asset went in the opposite direction – people began to think the probability was far less than 20 per cent. By the end of trading the price had collapsed to only $0.001. The market's judgement was good as the film proved a huge success. The actual ticket sales soared far beyond the specified range of $90 million to $100 million and ultimately exceeded $153 million.[23]

Studies of prediction markets suggest they often produce accurate forecasts, even when they don't involve real money.[24] Some companies now use the method to forecast whether cutting-edge research and development projects or new products will be successful. A market that allows Hewlett Packard managers to buy and sell assets has led to more accurate product demand forecasts than those obtained directly from the company's sales experts.[25] Forecasts from prediction markets have also outperformed other methods when predicting Oscar winners, box office successes, sporting champions and macroeconomic variables, such as levels of business confidence or retail sales.[26]

The *New Yorker* columnist James Surowiecki is a strong advocate of prediction markets and has argued their case in *The Wisdom of Crowds*. But, like all methods, they have their disadvantages. Unlike well-run Delphi applications or supermeetings, people don't share information or rationales. The only indication that we have about what other people are thinking is the current price. And should that price suddenly move we might think they know something that we don't. So if a price falls, even slightly because of a random blip, we might sell. This lowers the price further so other people sell. The results can be a cascade – a big change in the price for no reason other than the fact that the market is behaving like a jittery herd of sheep. Coupled with this is the danger of what is known as favourite-longshot bias. This describes the tendency, observed in horse racing, that people have for underrating the chances of favourites winning and overrating the chances of a longshot. In prediction markets this often means that the market's estimate of the probability of highly unlikely events will tend to be set too high.

Also, we argued that a key advantage of prediction markets is their incentives – we can make money from our forecasts so it pays to think hard about them. While this might work in the short term it will be less of a motivator in long-term forecasting. Suppose that each asset will pay out $1 if live bacteria are found on Mars by 31 December 2030. We could have a long time to wait for our money.

All in all, we have seen that groups can veer between incompetence and brilliance when they set out to anticipate the future. But brilliance is unlikely to occur naturally, so groups need careful handling. We shouldn't trust a forecast from a comfortably collegial and confident group whose members pontificate around a table supporting each other's positions. But forecasts from groups who debate and challenge and show some humility in the face of uncertainty are likely to be more worthy of trust. If this looks impractical and we suspect that a group will sag towards conformity like a leaking balloon, then methods like Delphi or prediction markets may be needed. That way we can probably put our trust in the uncanny wisdom of the crowd.

# CHAPTER 6

# FORECASTING
# OURSELVES

**W**hat about forecasting our personal futures? These will probably be affected by macroeconomic swings like recessions or booms. If so, we'll be interested in knowing whether another economic collapse is predicted or whether a golden age awaits us. New technology and future discoveries, like driverless cars or cures for cancer, are also likely to impact on our futures. So, again, the latest forecasts of what's on the horizon may grab our attention.

But these are all big global events. Individually, we also make lots of forecasts about our personal future circumstances. What will it be like to drive the new car that's being delivered next month? Will I be happy in my new job? What will next summer's holiday in Greece be like? We don't usually extrapolate trend curves or calculate probabilities for these events – though there are doubtless some people who do – but we make forecasts implicitly. We may even run an imaginary scene in our heads – we find ourselves walking along a sun-baked empty beach on a Greek island, listening to waves lapping against the shore. Relaxed and happy we think what a great holiday this is – excellent food, a friendly hotel, lots to see and do. Will it really be like that? Are we good at forecasting how we will feel and how we'll react to future events, good or bad?

### Half-full glasses and tumblers

In 1993, a 28-year-old British woman looked back on her life and saw herself as a failure. Her marriage had ended after only thirteen months. She was a single parent with no job and was 'as poor as you could be without being homeless'. She'd always dreamt of being a novelist, but her parents had treated this as an amusing personal quirk. From a poor background themselves, they had urged her to follow a safe and secure career. She seemed to have little chance of being a successful novelist. After all, publishers are bombarded with manuscripts from aspiring writers and only a tiny fraction of these are published. Even fewer become bestsellers. In a country where the media is dominated by privately educated people who attended universities like Oxford and Cambridge, what chance did a woman from a modest background have?

But the woman persevered in pursuit of her dream. One day she found herself on a train that was delayed for four hours. During that long wait an idea came to her for a novel.

By the end of the journey the idea was fully formed, but it took two years to complete the writing. Twelve publishers then rejected her manuscript. Many people would have given up at this point but the woman, supported by her literary agent, clung to her belief that she would be successful. The book was eventually accepted by a publisher, but only because his eight-year-old daughter pleaded with him to print it.

That woman was J. K. Rowling and the manuscript contained her first Harry Potter novel. The series of books that followed has been the bestselling in history. Over 400 million copies have been sold and films based on the books have grossed millions of dollars. Rowling is now world-famous. But without the benefit of hindsight she appears to have been unrealistically optimistic.[1] Huge odds were stacked against her. Had she thought only about the typical chances of success she would surely not have persisted with her dream. It was Rowling's optimism that pushed her forward despite the many obstacles that stood in her way. Without that, she would have failed.

This almost delusional, positive belief in ultimate success, that some people possess, has made the world what it is today. History is full of stories of those who battled on despite little apparent chance of success and went on to achieve their goals. Winston Churchill once said: 'I am an optimist. It does not seem too much use being anything else.' When Britain faced an imminent invasion by Hitler's mighty Wehrmacht in 1940, many in the British establishment wanted to sue for peace.

Churchill chose to fight on. He inspired the nation to follow him. Five years later the Nazis were defeated. The Wright brothers endured numerous failed attempts before they succeeded in flying a plane on the beaches of North Carolina. Only a few years earlier, in 1895, the ever-confident Lord Kelvin, one of the most eminent scientists of his day, had declared that 'heavier-than-air flying machines are impossible'. The teachers of Ludwig van Beethoven considered him hopeless. They predicted that he had no chance of succeeding as a composer. We could go on and on…

A positive attitude can bring many other benefits in addition to promoting the self-belief people need to battle almost impossible odds. Olga Stavrova and Daniel Ehlebracht of the University of Cologne recently found that those with a negative, cynical outlook on life earn on average £2,000 less per year than others.[2] Apparently, they tend to forgo the opportunities and benefits of co-operating with other people, because they suspect the motives of potential collaborators. Their suspicions also mean they are reluctant to seek help when they need it. There's also evidence that optimism – even unrealistic optimism – can be beneficial to mental health. It can also lead to behaviour that's likely to promote health and well-being when a person is facing a serious illness.[3] Cardiac patients who thought their chances of suffering another 'cardiac event' were lower than those of a typical patient with their condition were less likely to see a repetition of the event in the subsequent twelve months.[4]

Indeed, unrealistic optimism may be essential to happiness. According to some researchers, there is a distinct group of people who are generally accurate at predicting future personal events. If you ask them to estimate their chances of suffering from particular illnesses or an early death these estimates will be pretty reliable.[5] These are people suffering from mild-to-moderate depression. The phenomenon has been called 'depressive realism'.[6] However, it's fair to say that the existence of depressive realism is not without controversy.[7] A recent review of seventy-five relevant studies found that the effect was small and depended on how the study was carried out.[8]

Nevertheless, some scientists even argue that our very survival as a species has depended on our brains being hard-wired for optimism. They have suggested that the hippocampus evolved to allow us to simulate possible futures. This gave us the evolutionary advantage of being able to plan and anticipate. However, this came at a cost. We could now envisage one certain and demoralising event in the future – our own death. Arijit Varki, a biologist at the University of California in San Diego, argues that the hopelessness arising from this realisation could have prevented us from performing the functions we needed to survive.[9] Without a mitigating mental mechanism the human race would have ceased to exist. That mechanism was unrealistic optimism. Its evolution, alongside our ability to imagine the future, it's argued, allowed us to avoid despair and to survive.

Despite all these benefits, unrealistic optimism is not good from a forecasting perspective. Decisions we make, based on an excessively sanguine view of the future, can do serious damage. For example, smokers may overestimate the chances they will be able to quit at some future date. So they carry on smoking, not realising that their habit is becoming more established.[10] When events do not turn out as well as we expected, we suffer disappointment and regret. When our performance in a sporting event or examination fails to live up to our expectations, we can suffer a loss of self-esteem. If we overestimate our chances of being able to pay off a loan for a new car or new kitchen we may find ourselves deep in debt. If we underestimate the risks of suffering from alcohol-related problems or disease, we may neglect to take the necessary precautions to prevent them from occurring. For example, people who were overly optimistic about their chances of not contracting the H1N1 swine flu virus indicated that they would wash their hands and use hand sanitisers less frequently than those who had a realistic perception of the risks.[11]

But, I hear you say, you argued that seemingly unrealistic optimism propelled J. K. Rowling to the top of the bestseller lists and brought heavier-than-air flight to the world. Surely, nothing great would be achieved if everyone gave up the fight when confronted with apparently hopeless odds. However, we must be careful not to be misled by so-called survivor bias. The truth is that there are thousands who have shared J. K. Rowling's dream and didn't succeed. These are people who may have toiled for years and made huge sacrifices to produce manuscripts, but never made it. Their efforts were simply dumped in the forlorn paper mountains that publishers refer to as slush piles before being sent back to their authors with a terse rejection note in a battered parcel. We only hear about the rare successes – the survivors. We hear about the tycoons who were born on the wrong side of the tracks but went on to make a billion. We hear about the inventors who prevailed despite mockery and rejection, going on to develop radar or the jet engine. The budding entrepreneurs whose businesses failed to take off, or the inventors who sunk their life savings into failed projects, are largely invisible.

However, this is *not* an argument against having dreams and ambitions. Nor is it an argument for giving up on these dreams. It is simply an argument for having a realistic forecast of one's chances of success. Whether we want to take that chance is a decision not a forecast. At least with a realistic forecast we'll know the risks we're taking. It's also important to emphasise that we've been discussing unrealistic optimism. As we'll see in Chapter 9, the act of making a forecast can change the state of the world. In some cases forecasting an event will increase its chances of happening. In market research surveys where people are asked to forecast their future purchases, it can increase the probability that they will follow through on what they have said, so the forecast becomes a self-fulfilling prophecy.[12] As with J. K. Rowling, when a person is optimistic it can influence their behaviour. It can motivate them to battle through difficulties, so in the end their optimism was not unrealistic after all.

## I'm different from the crowd

When it is unrealistic, optimism can show itself in two different ways. Sometimes we have a tendency to see favourable events in our lives, like maintaining our weight at a healthy level in the future, as being more probable than statistics suggest they are. Meanwhile, the chances of unfavourable events, such as contracting a sexually transmitted disease, are underestimated. But we also tend to believe that things are more likely to turn out better for us than for other people. In a seminal study, Neil D. Weinstein of Rutgers University found that college students believed they had a better chance than their classmates of getting a good job offer before graduation, of living past the age of eighty and of owning their own home.[13] In contrast, they thought they were less likely than their peers to suffer from a drinking problem in the future, have a heart attack before the age of forty or be injured in a car accident.

Why should this be? One reason is that we know much more about ourselves than other people. Author C. S. Lewis once wrote: 'We have inside information. We are in the know.'[14] This means that we focus on the specifics of our own plans, abilities and intentions when forecasting what will happen and, as we saw, we are often pre-programmed to have an over-optimistic view of our personal situation. This eclipses the underlying statistical data that may be available on the event we are forecasting. We pay little attention to these underlying statistical rates (or base rates). But when it comes to other people we know little about them. It's difficult to visualise a human being who is 'the average person with your age and background'. Hence for this abstract being we are more prepared to fall back on base rates when making our forecast. Ironically, this means that we are likely to make a more accurate forecast for these 'unknown' people than for ourselves.[15]

In some cases we replace this abstract being with a stereotypical image of the sort of person who might experience the event we are forecasting. For example, we might see the typical driver who is involved in a car crash as a young speed merchant who roars around town with his hi-fi booming, unaware of the needs of other road users. Few of us would see ourselves fitting this image so we assume our risk of being involved in a crash is lower.[16]

## A different tint

Despite these tendencies, we are not always wildly overoptimistic. In some situations we remove our rose-tinted spectacles, or at least change their lenses so they have a lighter hue. This happens when we see events as largely being beyond our control.[17] For example, suppose we're going abroad on vacation for a month and have to estimate the probability that our house will be burgled while we are away. Because we feel there's little we can do to protect our home while we are thousands of miles away we may come up with a more realistic figure for the probability. When we think we have control over events, we naturally perceive that we have the power to prevent negative things from happening. I can stay out of the sun to avoid skin cancer. I can use my chainsaw carefully to reduce my chances of an accident. This is the case even if the control we think we have is exaggerated or illusory. I can work harder in the office to reduce my chances of losing my job. If I choose my lottery numbers based on my children's birthdays, I'll increase my chances of winning. When we are aware that we don't have control over events we are more realistic.

Sometimes we go to the opposite extreme. We are unrealistically pessimistic. It's estimated that around one-third of Americans employ a tactic called defensive pessimism to cope with anxiety and worry.[18] Suppose you've been asked to give a public talk in six months' time. You were flattered to be asked, and an engagement that is in six months' time seems a long way off. Later you hear there's great interest in your topic. Over a thousand people will be attending. As the date approaches your anxiety increases. You feel your throat going dry at the mere thought of it. You envisage a scenario where you are standing at the microphone but your mind has gone blank. Not a word will come out of your mouth. The audience's murmuring turns into a cacophony of amusement. People start to walk out. 'What a waste of time that was!' someone shouts.

Then there's a scenario where you have that nightmare figure in the audience – the clever, condescending critic who has turned up to publicly humiliate you. They interrupt your talk and tear your arguments apart for reasons you just can't imagine. 'Heck, they're right,' you realise. You have no response. The rest of your talk is doomed. You are haunted by the smirk of triumph on the imaginary critic's face.

Having low expectations about future events enables defensive pessimists to anticipate threats. By expecting everything to go wrong they actively identify potential disasters and take steps to neutralise them. They might prepare a set of bullet points to hide below the lectern to give them prompts in case their brain stops functioning mid-sentence. They might work through their arguments again and triple-check that they are watertight and evidence-based. Even if things do go wrong they are better equipped to cope. They are braced for a loss so they are more likely to be pleasantly surprised by events than to suffer disappointment.

Defensive pessimists often suffer from low self-esteem.[19] But ironically, they also tend to be successful people because they are motivated to prepare carefully for future events. This is according to Julie K. Norem, a professor of psychology at Wellesley College who has written a book titled, *The Positive Power of Negative Thinking*.[20] Her research suggests that trying to dissuade people from employing defensive pessimism by encouraging them to think positively in fact negatively impacts their performance.[21] From a forecasting perspective defensive pessimism is really an attempt to exploit the curse of the self-destructive forecast. We make a forecast in the hope it will prevent what we are forecasting from happening. Or, in other words, we make our forecast with the earnest intention of getting it wrong.

## The psychological immune system

There's a line I remember hearing in a 1967 song by British musician Al Stewart: 'Sometimes I wonder how it feels to be Paul McCartney or the Queen.'[22] Many of us wonder what it's like to be known and admired throughout the world. Suppose that one bright morning you wake up with a beautiful original song in your head. You instantly write it down and then record a video of the song for YouTube. Your performance goes viral and within weeks you are an international celebrity. Top singers want to perform your song and beg you to write more. Streams of international awards cascade down on you from the music industry.

Can you forecast what this sudden fame would feel like? Presumably, there would be restrictions – perhaps you couldn't go out in public without hiding behind sunglasses – but on balance surely it would be great. You'd probably be wealthy, you'd get invitations to all sorts of exclusive social events and, as long as you managed to avoid envious attacks by the tabloid press, your ego would be constantly massaged. For years to come you would be on an emotional high.

Suppose instead that, while relaxing early one evening, you hear your mobile phone signalling that a text message has arrived. You expect that it's your spouse telling you they're working late. But it's from your boss. 'Sorry to let you know this way. But we've decided to dispense with your services with immediate effect. Your personal possessions can be collected from the security gate.' Can you forecast how you would feel afterwards? You'd probably expect to be devastated. Not only would you be angry about the way you'd been fired, but the sudden loss of your livelihood would wreck your self-confidence and leave you in a depressed emotional state for a very long time.

Neither of these predictions is likely to be accurate. Psychologists tell us we are usually wrong in our forecasts both of the immediate impacts of events and of their lasting effect on our emotions.[23] When forecasting the initial impact we tend to envisage extremes of happiness or misery that usually aren't borne out by the actual experience. Our predictions about which events are likely to be enjoyable or unpleasant are generally accurate. We are usually right when we anticipate that root canal surgery will be less pleasant than a sunny day lazing on the beach in Barbados. Where our judgements are askew is in our assessment of the intensity and the duration of emotional reaction to events. Winning the lottery or getting our dream job won't make us quite as happy as we imagine. And the feel-good effect will wear off faster than we think. Conversely, failing to get a promotion at work or splitting up with one's romantic partner probably won't make us feel quite as bad as we expect and we are likely to recover from the disappointment faster than we can imagine.

Making overly extreme forecasts of the intensity and duration of our reaction to events is called impact bias by psychologists.

We appear to over-forecast the intensity of our emotional response for two reasons. First, when we think ahead we imagine that our thoughts at the time will be dominated by the event to the exclusion of all else – every waking moment I will be thinking about the unfairness of being turned down for promotion. In reality, many other things will be going on in our lives in the future – some good and some bad – and these will serve to distract us from an exclusive focus on a negative event. A medical test we've been worrying about indicates that we're healthy; we win a prize in a local newspaper competition; our favourite football team wins the Cup and the central heating system at home breaks down. As a result of all these goings-on, the impact of the event is diluted. We don't feel quite as bad about the disappointment as we expected to feel.

When an event will lead to a permanent change in certain aspects of our lives, we tend to exaggerate the importance of these aspects while we downplay the many things that will remain the same. For example, in one study the psychologists David Schkade and Daniel Kahneman found that people living in the Midwest of the USA expected Californians to be happier because of the warmer climate there.[24] But for people living in California the weather is only one of many factors that determine their satisfaction with life. The researchers concluded that 'it is not unlikely that some people might actually move to California in the mistaken belief that this would make them happier … Nothing that you focus on will make as much difference as you think.'

Impact bias also occurs because we don't realise that our minds will set about making sense of events that are unexpected or new. Immediately after receiving the news that a partner has been cheating, we might feel shocked, puzzled and upset. But our mind will quickly get to work searching for reasons. They have revealed themselves to be untrustworthy. Their sister behaved the same way. It's now obvious that cheating and dishonesty are family traits. We couldn't live with that. We're better off without this person in our life. We are extremely good at changing our perception of the world – ignoring, distorting, rearranging or inventing information – to protect ourselves. In particular, we tend to believe that our successes are due to our skills, but our failures are caused by external factors.[25]

Once we feel that we've identified a rationale for what has happened it helps us to adapt to our new circumstances. The emotional rawness is reduced and we'll spend less time reflecting on the event. Eventually the shock appears to be less exceptional and we may even come to see it as having been inevitable. Sense-making has the opposite effect when we experience positive events. Our immediate joy on receiving surprisingly good news, like winning a prize or gaining an unexpected top grade in an exam, is soon tempered. As with negative events, the good news soon becomes woven into the everyday fabric of events. Our excitement subsides faster than we expect.

It's often said that time heals. This natural tendency of our minds to moderate our emotional reaction to negative events has been called 'the psychological immune system' by some researchers.[26] Like our physical immune system it serves to protect us from attacks. It allows our sense of well-being to return quickly to a relatively stable level in the face of the slings and arrows of outrageous fortune. Researchers at the University of Illinois at Urbana-Champaign and Indiana University at South Bend tracked the reactions of 115 people to negative events in their lives, such as the end of a romantic relationship, difficulties in getting along with colleagues, financial problems and the death of a close friend. They found that their subjective sense of well-being returned to normal surprisingly quickly.[27] The impact of most events had diminished within three months.

Why are we unaware that our psychological immune system will protect us from overwhelming melancholy? Why don't we realise it will work hard to restore us quickly to emotional normality when bad things occur? After all, we can often see its rationalising mechanism at work in friends and colleagues. John claims he failed his law examination because the questions were unfairly difficult. Mary says she didn't pass her driving test because the examiner took a dislike to her. One explanation is that the immune system wouldn't function, if we were aware of our ability to distort the truth to protect our self-esteem and well-being.[28] We have to believe in what we are telling ourselves. If we know that in the face of a future negative event we will attempt to comfort ourselves with porkies, then we won't be comforted at all. Also, if we are aware that we'll be able to explain away whatever comes to pass and incur minimal heartache, then our incentive to avoid negative events will be reduced.

While there are advantages to being oblivious to our flair for softening future blows, our ability to do so comes at a cost. Our forecasts will be wrong. And wrongly anticipating the effect of events – good or bad – can matter. Decisions about marriage, divorce, having an operation, giving a home to a pet, changing jobs, retiring early, writing a book and having children can all depend on forecasts of what we think it will be like once we have made the decision. We might resist having elective surgery that would benefit us if we overestimate the time it will take to recover emotionally from the operation. We might rush into marriage wrongly, anticipating we will be on an emotional high for the rest of our lives. Or we might decide not to apply for a great job because we overestimate the pain that will follow any rejection.

## A Lamborghini would change my life

It's not only wrongly predicting the effects of events that can lead to decisions we regret. When we decide to purchase a product we also implicitly forecast what the experience of owning the product will be like. And, for the same reasons, we often get this wrong too. For products we haven't yet tried we neglect to forecast that we will adapt to owning the product. Possessing it will soon become an ordinary and everyday experience. As with the addiction to a drug, further purchases of new products will be needed to give our well-being another temporary boost. I spend half of my month's salary on a top-of-the-range high-definition television with lots of extra features. A few weeks later, and by the time the bill appears on my credit card statement, it doesn't seem anything special.

Paradoxically, once consumers have had some initial experience of owning a product they do anticipate that their enjoyment of it will decline. But this time they overestimate the speed with which this will occur.[29] They expect the novelty to wear off faster than it really will.

Can we help people to make better forecasts of their reactions to events or products that they buy? Several strategies seem to work. Getting them to think about a wider range of things which might occur in the future reduces their focus on the event in question. This helps them to appreciate that this one event won't dominate their future existence. They then begin to realise that their emotional response to the event will be diffused and forecast a less extreme reaction to it. People can also be reminded of what their original forecasts were and shown that these were too extreme. This makes them less likely to come up with an overly drastic forecast in similar circumstances in the future.[30] For example, we could record our expectations and fears in a diary and then look back to see if these were borne out. This could help us to realise that things won't seem quite so bad (or good) as we might otherwise have predicted.

## Forecasting our own performance

There are times when we find ourselves trying to anticipate how well we'll perform on a task or a test. How well will I do in the history examination I'm due to take next week? Will I pass or fail? Do I have any chance of getting a distinction? How will I fare compared to the rest of the candidates? Will I get an above average grade? Might I even make the top 10 per cent? Forecasting our future performance can be important. If we underestimate our ability then we might forgo opportunities where we would have excelled. We could turn down the opportunity of a job where we would have been a great success because we wrongly doubt our ability to perform well. If we over-forecast our future performance we might refuse help or advice, thinking that we don't need it.

So how good are people at forecasting how well they will perform on tasks or tests? A common finding is that high performers tend to underestimate their future scores. In contrast low performers tend to overestimate how well they will do. There's some controversy as to why this happens. One of the most widely discussed explanations is the Dunning-Kruger effect,[31] named after two psychologists, David Dunning and Justin Kruger, of Cornell University who proposed their theory in 1999. This postulates that relatively unskilled people are unaware of their lack of ability. English poet Matthew Prior wrote: 'From ignorance our comfort flows.' Relatively poor performers, according to the theory, live with the comfortable illusion that they are better than they really are. On the other hand, high performers find tasks easy and erroneously assume this is true for everyone else. Hence they tend to assess their ability as being close to the middle range and make forecasts that are consistent with this.

There is plenty of evidence that people who lack relevant skills tend to overestimate their ability to perform well on tasks. In one study 80 per cent of drivers considered themselves more skilful than the average driver, suggesting that there are many who are unaware of their relative lack of proficiency behind the wheel.[32] The same is true of people's perceptions of their ability at football, business management and leadership.[33] In their paper, Dunning and Kruger relate the story of McArthur Wheeler, an incompetent criminal. One morning in 1995, the 44-year-old decided to rob two banks in Pittsburgh without wearing a disguise. His smiling face was recorded on CCTV cameras as he pointed a gun at a teller demanding cash. That night the surveillance tape was played on the eleven o'clock TV news. By the end of the day Wheeler had been arrested after a viewer contacted the police with his name. When the police played the videotape back to him he couldn't believe what he was seeing. Before the robberies he had covered his face in lemon juice. He knew that lemon juice could be used as invisible ink and he'd assumed that it would also make his face invisible. 'But I wore the juice,' he is alleged to have mumbled.

As the *New York Times* put it: 'If Wheeler was too stupid to be a bank robber, perhaps he was also too stupid to know that he was too stupid to be a bank robber.' Whatever, the cause, Wheeler's overly optimistic forecast that he would get away with the robberies cost him a lengthy jail sentence.

Dunning and Kruger's explanation of why the unskilled are overly optimistic about their performance and the skilled overly negative has been cited by other researchers thousands of times. It even won them the Ig® Nobel Prize in Psychology in 2000. The prizes are awarded for 'achievements that first make people laugh then make them think'. However, simpler explanations for the finding have also been put forward.

I collaborated with two colleagues at the University of Bath in a study where students were asked to forecast the marks they would achieve on a series of in-class tests.[34] The students' forecasts were correlated with the marks they actually obtained, suggesting that they did have some insights into how well they would perform. But like other researchers we found that those achieving high marks tended to forecast too low a result, while the low scorers were overly optimistic. There was evidence that the students would first estimate what the average mark for the whole class would be. To produce a forecast of their own mark they would then either increase or decrease this figure based on their assessment of their own ability. But their estimate of the average mark for the whole class appeared to act as an anchor. As Chapter 2 showed, people tend to stay too close to anchors when making estimates. As a result the top students did not increase their predicted marks sufficiently above the average to reflect their true performance so they under-predicted how well they would do. Similarly, the students who struggled on the test made downward adjustments from the average that were too small and so they overestimated their performance.

Of course, the true explanation for why we misjudge our abilities may involve a combination of the Dunning-Kruger effect, anchoring and other factors. Whatever the cause, it's perhaps surprising how often we get predictions about ourselves wrong. In the forecourt of the Temple of Apollo at Delphi – the sacred place where the Ancient Greeks travelled to hear forecasts about their future – it is claimed that the following aphorism was inscribed: 'Know thyself'. But knowing thyself is clearly not easy. Benjamin Franklin wrote: 'There are three things extremely hard: steel, a diamond and to know one's self.' The fact that it's hard to know ourselves means the predictions we make about our future reactions to events are too extreme. We are unaware of the extent that we will adapt to new circumstances. We misjudge how much we will enjoy products that we buy and how long our enjoyment will last. And we misjudge our abilities. So the forecasts we make about how we will tackle tasks and tests are usually askew.

# CHAPTER 7

# I SPY IMPOSTORS!

**Agency problems**

Have you ever suspected that a garage is trying to persuade you to have unnecessary repair work carried out on your car? I was once told that my car needed a new clutch and, by a happy coincidence, the garage just happened to be running a special deal on clutches. The clutch seemed fine to me so I rejected the generous offer and drove the car for another 60,000 miles with no problems. Or perhaps an estate agent has encouraged you to sell your house at a reduced price by wrongly telling you the market for houses like yours is going through a slump. The quicker your house sells, the faster they'll get their commission. Or maybe a travel agent has booked you on a more expensive flight to New York because they receive a higher commission from the airline they've chosen.

In all these cases we have what is known as an agency problem.[1] We are paying a person or an organisation to act on our behalf. The agent has more expertise or information than us, but they are using this to further their own interests, not ours. Can the same problem occur in forecasting? You bet it can. Often forecasters are tempted to produce predictions that they think will benefit themselves rather than genuinely indicating what they think the future might hold.

Several years ago, I was researching how accurate managers' judgemental forecasts were compared to forecasts from computers. It was the early days of the internet and I obtained a spreadsheet from an internet service provider containing a record of managers' forecasts of the numbers of new customers they would sign up each week. The forecasts looked awful. Every week they had underestimated the actual number of customers and in most cases the differences were substantial. In one week the number of new customers was nearly double what had been forecast.[2]

The forecasts were so bad that I assumed there must have been some mistake in the spreadsheet and phoned my contact in the company.

'No, what I sent you is correct,' he replied. 'I forgot to tell you that the forecasts are produced by the marketing department and they look good if they exceed the forecast.' By deliberately producing an incorrect forecast, staff could then go to their senior managers and say, 'Hey we've beaten the forecast again; look at what a great job we're doing.'

Unfortunately, the staff in the operations department were less impressed. They used the forecasts to plan workloads for the following week and needed to have an idea of how many welcome packs to send out to new customers. Each week they were caught out by the higher than expected demand. The operations staff had every right to be annoyed – the information they were being given was not a forecast at all. It was a decision. The marketing department were not trying to make an honest estimate of what the future customer numbers would look like. Instead, they were choosing a number that they judged would maximise the kudos they received. Presumably, the extent of the deliberate underestimation was constrained by the need for the 'forecasts' to have some semblance of credibility. The underestimation decreased as the weeks went on, suggesting that complaints from the operations staff were having some effect. But forecasting should not be about balancing looking good against the risk of being found out.

The desire to look good, or at least to avoid looking bad, can have other effects. As forecasters receive new information which is likely to have implications for the future, it's rational for them to revise their original forecasts. The problem is that, if a forecaster keeps changing their forecast, people might begin to see it as a sign of incompetence. This means there is often an incentive to keep a forecast the same despite the new information, or at least to make only small changes. Evidence for this reluctance to make big changes was found in a study of macroeconomic forecasts made by the Association of German Economic Research Institutes over a period of thirty-five years.[3] The study indicated that the tendency was particularly pronounced for forecasts that were subject to a lot of media coverage, such as national income forecasts. The association may have feared that making large changes to a forecast would be seen as an admission that a gross mistake had been made. Sometimes people using forecasts put pressure on forecasters to keep them the same. President Jimmy Carter allegedly complained about the inconsistency of forecasts by his economic advisors and hinted that he would be better off using a fortune-teller at the Georgia State Fair.[4]

Forecasts involving probabilities can suffer from the same biases. Gideon Keren recounts an experiment where people were told it had rained on three out of four days.[5] They were then shown the probability forecasts of two different weather forecasters for these days and asked who they thought was the more accurate forecaster. Forecaster A's forecasts of rain for the days were: 90 per cent, 90 per cent, 90 per cent and 90 per cent. Forecaster B's were 75 per cent, 75 per cent, 75 per cent and 75 per cent. Clearly, by making the same forecast for all four days, neither forecaster had done a good job of discriminating between the days when it would rain and those when it wouldn't. However, Forecaster B was better calibrated – it had rained on 75 per cent of days when they forecast rain (we'll look at calibration again in Chapter 10). Despite this, almost two thirds of those asked preferred Forecaster A. They thought that B's forecasts lacked decisiveness. People have a preference for forecasts that give probabilities which are closer to certainty. It seems more extreme probabilities are associated in their minds with forecasters who are more knowledgeable and accurate, irrespective of the true quality of their forecasts.[6] The temptations for a forecaster are obvious. They might honestly believe an event only has a 65 per cent probability of happening, but reason that giving a forecast of 90 per cent is more likely to get their forecast heard and adopted. Quoting 65 per cent could be seen as a sign of incompetence.

### Stand out from the crowd or stick with them?

Forecasters face many other temptations to change their forecasts from what they honestly believe will happen. Economic and financial forecasters often seek to enhance or preserve their reputations by basing their forecasts on what other experts are predicting.[7] But some forecasters deliberately make their forecasts unique, a practice known as contrarianism.[8] They do so in the hope this will make them stand out from the crowd. If they're lucky and theirs is the only forecast that proves correct, this will attract publicity. Being wrong probably won't matter because people can be surprisingly quick to forget. Some potential clients who are searching for a 'good' forecaster may base their choice on the accuracy of this one forecast and ignore, or be unaware of, a relatively poor past record that has preceded it. As George B. Henry stated in 1989 when he wondered whether economists were 'worth their salt':

> '... one or two strikingly unorthodox predictions that prove accurate can make a career ... If you're hot, you'll get

favourable publicity and so will your firm. And, during those periods when you're consistently wrong, so what? You'll surely have plenty of company, and being right or wrong does not seem to matter … after you appear in the press a few times, you become an authority figure in customers' minds."[9]

One study[10] investigated forecasts of exchange rates of a range of currencies with the US dollar made by professional forecasters working at institutions such as investment banks, research institutes and universities. It found significant evidence of contrarianism. A similar result was found when the same researchers studied professional forecasters' predictions of oil prices[11] and metal prices.[12] Other researchers[13] found that US macroeconomic forecasters sought to differentiate their forecasts when predicting variables like economic growth, inflation and unemployment.

Deliberately making a forecast different to those of others also acts as a signal that a forecaster is confident in his or her ability.[14] This probably explains why older American economic forecasters were found in one study to produce bolder forecasts than their younger counterparts.[15] Once they feel that their reputation has been established there is less risk involved in 'sticking your neck out'. Unfortunately, for those buying the predictions of older forecasters, it came at a cost. As forecasters became more senior, more established and, presumably, more confident in their ability, their pay went up but their accuracy worsened.

A broken clock is correct twice every day, and one simple strategy for keeping our forecasts different from others most of the time is to carry on forecasting the same thing. Forecast a recession at the start of every year and eventually one will come along and then we'll be able to trumpet our forecasting accuracy. Dr A. Gary Shilling, a well-known American financial analyst, has been accused of following exactly this 'broken clock' strategy.[16] For many years he has forecast recessions and stock market crashes.[17]

Shilling's website highlights his correct predictions of a recession in 1969, 1973 and 1991. But as a writer on *ValueWalk* (a news site) pointed out, weather forecasters wouldn't get paid if they forecast rain every day and lauded their achievements on the days when they were right. Perhaps they would get paid if they were able to hide the times when they'd been wrong. In 1987 the Wall Street index claimed that A. Gary Shilling & Co. had mailed clients a copy of an article that had appeared in the *Wall Street Journal*. One paragraph showed that, in a survey of forecasters, Dr Shilling had made the best forecasts regarding thirty-year treasury bonds. Oddly, according to the financial researcher Owen Lamont, another paragraph of the article had been omitted.[18] The missing paragraph had stated that Dr Shilling had tied for last place in his bond forecast made six months earlier.

Forecasts that stand out can sell newspapers, and in Britain, newspapers like the *Daily Star* and the *Daily Express* often print startling headlines about imminent weather conditions.

'Britain to bake in SIX-week heatwave. RECORD SUMMER ON WAY', shouted the *Star*'s front page on 19 May 2014. But in a review of summer 2014 the UK's Meteorological Office reported: 'Summer 2014 saw several spells of fine, settled weather in both June and July but no major heatwaves'.[19]

'100 DAYS OF HEAVY SNOW: Britain now facing worst winter in SIXTY YEARS warn forecasters,' yelled an *Express* headline in November 2013. This was followed by one of the mildest winters on record with no snow at all in most places. Up until 2012 many of the stories in the *Express* were based on forecasts by Positive Weather Solutions, a company run by a man named Jonathan Powell. The newspaper was happy to feature pictures of the company's forecasters alongside its reports – attractive young women who possessed what writer George Monbiot called 'prominent credentials'.[20] When Monbiot did a picture search of the forecasters he found that they were amazingly versatile and busy. One was also working as a mail-order bride, a hot Russian date and a hot Ukrainian date, while another was able to offer services 'as an egg donor, a hot date, a sublet property broker in Sweden, a lawyer, an expert on snoring, eyebrow threading, safe sex, green cleaning products, spanking and air purification'.[21] The pictures were, in fact, stock photographs. When Monbiot asked Powell for the phone numbers of his forecasters he said he couldn't find them and promptly closed the company.

But that wasn't the end of it. A few months later the *Express* was reporting yet more extreme forecasts from Powell's new company, Vantage Weather Services. On 5 May 2012 it forecast that winter would 'last until June'. Yet on 23 May the *Express* subsequently reported that Britain was 'Hotter than Africa as Scotland is set to hit 79F'. Vantage Weather Services also now seems to have closed, though a picture of a pretty woman still appears on its website. As Monbiot pointed out, the newspaper and Powell enjoyed a symbiotic relationship. His forecasts helped the newspaper's circulation; he gained from the publicity. By the time last week's newspaper was in a recycling box most of its readers had probably forgotten all about the screaming headline.

In 1936 John Maynard Keynes wrote that 'worldly wisdom teaches that it is better for reputation to fail conventionally than to succeed unconventionally'.[22] Keynes's quote could also act as motivation for the opposite of standing out from the crowd: herding. A forecaster who herds goes along with what the majority of other forecasters are predicting, despite what they genuinely think the future will hold. Being with the herd is safe – if the forecast turns out to be badly wrong, they won't be singled out for opprobrium and their career is less likely to be damaged. Not surprisingly, herding is more common among relatively inexperienced forecasters. One study found that younger security analysts tended to produce their forecasts of company earnings after those of older analysts had been announced and their forecasts then tended to be closer to the consensus.[23] Another study of people making interest rate forecasts found that herding becomes more prevalent when there is more uncertainty about the future.[24]

Of course, what may appear to be herding could instead be a rational incorporation of other people's views into one's own forecasts. These views may be a legitimate source of information and can be used alongside other data to help a forecaster form a more accurate view of the future. Equally, a group of forecasters may all have access to the same information and thus form similar views. In addition, some academic studies that have found evidence of herding may be flawed because the statistical analysis they used gave misleading results.[25] But where a forecaster is herding in order to protect their career or their reputation, or practising contrarianism to attract attention, they are decision-making, not forecasting.

## Political pressures

Politics and forecasting can be a toxic combination. The Scottish poet and novelist Andrew Lang said in a 1910 speech that politicians use statistics in the same way that a drunk uses lamp-posts: for support rather than illumination. The same may be true of forecasts. And political interference in forecasting is rife in all sorts of organisations.

In a survey I jointly conducted of mainly American company forecasters, 52 per cent of the respondents indicated that senior management changed their forecasts, and in more than a quarter of these cases they altered the forecasts without even consulting the forecaster.[26] While these changes might have been made because the senior manager had knowledge of future events, which they hadn't shared with the forecaster, this seems unlikely, particularly as there was no consultation with the initial forecaster. Altering the forecast to make it more politically palatable was the likelier motivation.[27] For example, in another study, Craig Galbraith and Gregory Merrill[28] found that senior management in organisations frequently requested that forecasts should be changed to levels that were politically more favourable. Sometimes the favourable level was determined in advance and the forecast then had to be generated to meet this level. In some cases the statistical model was manipulated until it produced the desired forecast so it appeared that the forecast had been generated objectively – as in the pharmaceutical company we encountered earlier. The most common motive for changing the forecast was to influence internal resource allocation, for example to get more staff or a bigger budget for the department making the forecast. But these machinations often came at a cost: in 60 per cent of companies that were surveyed they tended to reduce forecast accuracy; only 15 per cent reported that accuracy was generally improved.

Of course, it's not only companies that mix forecasting with politics. When analysing forecasts made by US state governments of the revenue they would receive from sales tax, Stuart Bretschneider and Wilpen Gorr found there were biases dependent on whether the state had a conservative or liberal government.[29] For example, in times when money was becoming tight, conservative administrations tended to produce overly optimistic forecasts of tax revenues to avoid calls for tax increases. Only when monetary problems became serious and public did they switch to the opposite bias of underestimating revenue so that they could then use this as a rationale for expenditure cuts. At least one state has recognised these problems: in 1997 Washington state created an independent forecasting agency to produce 'unbiased, accurate and understandable forecasts of the demand for the major state services'.[30] The result: forecasting accuracy improved.

International politics can also cast its shadow over forecasting. The International Monetary Fund (IMF) regularly produces economic forecasts for many developed and developing countries, including forecasts of economic growth and inflation. Initially, these forecasts are based on sophisticated statistical models, but these are often adjusted. When researchers investigated adjustments for the period 1999 to 2005 they uncovered clear evidence of political bias.[31] The USA is the major shareholder in the IMF and, with about 17 per cent of the total votes, it can effectively veto major decisions, including who gets the Fund's managing director job. Countries that voted with the United States in the United Nations General Assembly were found to be rewarded with lower inflation forecasts when their governments were close to domestic elections. In many cases, the growth forecasts were also conveniently positively biased.

Inflation forecasts were also lower when a country had borrowed heavily from the IMF. It's in the IMF's interests to project a reputation that it is a competent financial institution and forecasting high inflation for these countries would reflect badly. For some heavily indebted countries there is even a danger that they will default on repayments which would also damage the IMF's reputation. Further loans can avert immediate dangers of default, but these need to be justified. Lower inflation forecasts just happen to be helpful in supporting such justifications.

## Scary forecasts

Many forecasts are deliberately intended to scare us into doing all we can to avoid some disaster so that what is being forecast won't happen. We are told that it's probable that more than half the UK population will be obese by 2050,[32] that global temperatures could rise by up to 3°C by 2050[33] and that child poverty levels will reach 24 per cent in the UK by 2020 if current government policies continue.[34] Sometimes these forecasts are produced with the best of intentions and when they are honest extrapolations of past trends based on the assumption that no action will be taken to reverse these trends, they can be regarded as useful forecasts. As Nobel laureate F. Sherwood Rowland argued when discussing climate change: 'What is the use of having developed a science well enough to make predictions if, in the end, all we are willing to do is stand around and wait for them to come true?' However, so-called forecasts cease to be forecasts when they've been deliberately manipulated to serve the political interests or other objectives of the people producing them.

Forecasts of future global temperatures produced by the Intergovernmental Panel on Climate Change (IPCC) can be particularly scary, but they have fuelled a fierce battle between the majority of climate scientists and an assortment of global warming sceptics. The intensity of the conflict means there's a danger of forecasts being adapted to support the arguments of those producing them. The danger is exacerbated by the huge amount of uncertainty associated with the forecasts produced by the IPCC. Different scenarios lead to different projections. For example, if greenhouse gas emissions continue at current levels, several ranges are given for probable mean increases in surface temperatures in the period 2081 to 2100, compared to 1986 to 2005, such as 0.3°C to 1.7°C or 2.6°C to 4.8°C.[35] These seemingly small differences in temperatures have major implications. For example, an increase of 4°C to 5°C would see humans deserting southern Europe, North Africa, the Middle East and other sub-tropical areas because of heat and drought. They would have to move towards the poles where all sea ice would have melted.[36]

There are many reasons for this uncertainty. The models used by climate scientists to simulate changes in the Earth's atmosphere and oceans try to specify how different physical processes behave and interact. But some of these processes, such as the effects of clouds and the formation of water vapour, are not well understood.[37] Also, large increases in carbon dioxide levels take us into uncharted territory. For example, scientists cannot reliably estimate the capacity of forests to absorb carbon dioxide if it exists at twice current levels.

In addition, there are controversies about how temperatures should be measured and uncertainty about how people will behave as temperatures rise. There is even the potential for errors in the computer programs that are used to calculate the forecasts, given the complexity of the models.

This uncertainty has helped influential people like US President Donald Trump, former President George W. Bush, ex-Australian Prime Minister John Howard and Britain's former Chancellor of the Exchequer Nigel Lawson to reject forecasts of global warming. One prominent forecaster who supports their position is Scott Armstrong, a marketing professor at the Wharton School of the University of Pennsylvania whom we've met in earlier chapters. Armstrong and co-researchers Kesten Green of the University of South Australia and Willie Soon of the Harvard-Smithsonian Center for Astrophysics have asserted that the 'best' forecast of future global temperatures is a simple 'no change' forecast – mean temperatures in the future will be the same as they are today.

In 2007 Armstrong challenged former US Vice-President Al Gore, now a well-known climate change activist and author of *An Inconvenient Truth*, to a $10,000 bet about changes in temperature levels over the succeeding ten years. Armstrong's 'no change' forecast would use the most recent year's average temperature at each of ten geographically dispersed weather stations as the forecast for each of the years in the future. Gore was invited to use the forecasts of any 'currently available fully disclosed climate model'. He didn't take up the challenge.

Despite this, Armstrong's 'Climate Bet' website[38] regularly reports on whether he or Gore would be ahead based on the forecasts made so far. Headlines like: 'May 2015: Now twenty-eight months straight of surprisingly low temperatures for Mr. Gore' regularly claim that the 'no change' forecasts are currently in the lead. However, the entire basis of the bet was questioned by people like climatologist Gavin Schmidt, director of the NASA Goddard Institute for Space Studies in New York. He described the challenge to Gore as simply a bet on the year-to-year random movements of the weather, rather than a bet on climate change. With huge uncertainties associated with climate forecasts it seems inappropriate to have a bet based on single figure forecasts. This is especially the case since ten years is, in climate terms, a relatively short period – too short to draw valid scientific conclusions according to a recent report.[39] As we have already seen, the IPCC forecasts do attempt to embrace uncertainty by providing multiple forecasts based on different scenarios and different models and each of these forecasts are expressed in their reports as ranges. Comparing multiple ranges from a climate model with a single number forecast looks like an unbalanced contest – apples and oranges come to mind.

Whatever the merits of the bet, both the climatologists, and some of the global warming sceptics, have been accused of slanting their forecasts to serve their own interests. The climatologists have been accused of having a career interest in promulgating predictions of global warming. Michael

Crichton's novel *State of Fear* chimes with this theme. It describes a world where climate science is being used to induce fear in citizens to keep them in order following the demise of the threat of communism. In other books and blogs, accusations are made that scientists are allegedly being pressured into toeing the global warming line to secure research funding, or to avoid dismissal if they espouse heretical views.[40] The Heartland Institute, a libertarian and conservative US-based think tank that rejects forecasts of global warming, stated that one of its plans for 2015 was to 'defend and recognize **Dr. Willie Soon** [emphasis in source] and other brave scientists who challenge unscientific claims that global warming is a crisis'.

On the other side of the argument many climate change sceptics have been accused of representing the interests of industries, such as those producing fossil fuels. In February 2015 the *New York Times* carried an article which alleged that when publishing academic papers on his research, Willie Soon had failed to disclose he had received $1.2 million from the fossil fuel industry over the previous decade, including at least $409,000 in funding from a subsidiary of the Atlanta-based Southern Company.[41] Southern has huge investments in coal-burning power plants. For several years, according to the article, it has spent significant sums lobbying in Washington against greenhouse gas regulations. One of the papers referenced by the newspaper was co-authored by Soon and appeared in the *International Journal of Forecasting* in 2009 advocating 'no change' forecasts.[42] An organisation called the

Climate Change Investigation Center, based in Alexandria, Virginia, accused Soon of a conflict of interest, alleging that he had promised the paper as a 'deliverable' to the Southern Company in exchange for their funding without declaring a possible conflict of interest to the journal. One of Soon's co-authors, Scott Armstrong, fiercely refuted the allegations.[43] The arguments roll on...

To risk a pun, environments where there are vested interests or agendas are unlikely to provide a healthy climate for forecasts which should be honest expectations of what will happen in the future, preferably with acknowledgements of uncertainty. They will, however, provide a healthy climate for decisions – numbers deliberately chosen because they support someone's case or interests. And in reality they should be called just that – decisions not forecasts.

# CHAPTER 8

# THE LURE OF
# ONE NUMBER

As a mature student in the early '80s I spent a summer at an English cider company working on one of their forecasting problems as part of my course. The company's draught cider was sent out to pubs and clubs in stainless steel kegs. Eventually, those that weren't lost came back empty to the cider mill where they were washed and reused. Sales of cider were booming and the managers were terrified that one day they would run out of empty kegs and have to dump thousands of gallons of cider. But new kegs were expensive so they were reluctant to buy them if the current number of kegs was sufficient to meet their needs. My job was to forecast how many empty kegs would be returned each week so that any shortfall could be anticipated.

The managers had come up with the idea of painting a hundred kegs in psychedelic colours and numbering them so that, when one of these kegs arrived back empty, staff in the washing bay would notice it. They could then record the time it spent between going out to a customer full of cider and its return. If you knew these typical times, the managers surmised, then you could make a prediction of the number of kegs coming back each week based on the number that went out in earlier weeks. But their idea had a fatal flaw: the painted kegs were so attractive that every single one of them was stolen while they sat outside pubs awaiting collection – hence my arrival.

I spent several weeks number crunching on 1980s computer technology, searching for patterns that would allow me to devise a method for making the forecasts. Finally, the stressful day arrived when I was to present my forecasting model to the managers. They seemed to like the method I had devised, but then one of them said: 'I can get this week's figure. Tell me your forecast for this week and then we'll see how good your method is.'

I started listing a few probabilities: 'there's a 10 per cent chance that you'll get less than 1,200 kegs back, a 15 per cent chance that you'll receive between 1,200 and 1,600...' but he interrupted.

'That's too much information. Just give me a single number.' I took a look at my model and quickly worked out the most probable outcome: '2,132 kegs,' I said.

The manager phoned his secretary to get the actual figure. In two long minutes it felt as if a whole summer's work, and indeed my entire future career, were hanging on one number. I felt my heart pounding.

'Two thousand, four hundred and fifty-one,' he repeated back to her as he wrote the number down. 'Hmm not bad, I suppose. About three hundred kegs out.' He didn't seem too impressed.

## Very abbreviated forecasts

We use forecasts when facing uncertainty, but single numbers tell us nothing about the level of uncertainty we are facing. To show uncertainty, forecasts can be expressed using probabilities and one organisation that has pioneered their use when presenting forecasts is the Bank of England. For example, in November 2016 it made a forecast of what the growth rate in GDP would be in the UK economy in quarter three of 2018, based on a set of assumptions about future interest rates and other policy measures. This is shown below.[1]

| Growth rate | Probability |
|---|---|
| Less than 0 | 19 per cent |
| 0 to under 1 | 19 per cent |
| 1 to under 2 | 22 per cent |
| 2 to under 3 | 19 per cent |
| 3 to under 4 | 13 per cent |
| 4 and over | 9 per cent |

This type of forecast is known as a probability distribution (note that the probabilities amount to slightly over 100 per cent because of rounding). We can see that the Bank acknowledges that it can't be certain about future growth, but it's pretty confident it will be below 3 per cent. In particular, it sees a significant risk (19 per cent) of negative growth. Forecasts like this, that convey risks or their absence, lead to better-informed decisions.

So why aren't all forecasts like this? Part of the problem lies with us as forecast users. We often resist having too much information to absorb, especially if we're busy, and as we'll see, sometimes we misunderstand the meaning of the probabilities. But deep down also lurks our innate dislike of uncertainty, so we often prefer to ignore uncertainties and turn them into apparent certainties.

Large sets of numbers tax our brains. Despite the billions of neurons in our heads we have a limited capacity to process information. For example, psychologist George Miller famously estimated that we can only hold 7+2 items of information in our working memory at any one time.[2] Constraints like these mean we tend to prefer single number summaries of information.

When studying the performance of schools or universities we usually just focus on their league table position rather than the underlying scores and their components. In reality the scores are often very close over a wide range of rankings so that minor random factors or measurement errors may account for any differences. But if our alma mater has just gone up in the rankings from number five to number three, that's all we want to know. Similarly, when choosing between products we sometimes decide on the basis of one criterion alone (e.g. which is the cheapest brand?) because combining information on price with information on after-sales service, quality and durability involves too much effort and time.[3]

Single numbers also look clear-cut and precise and even have an air of accuracy that can reflect well on the people producing them. But in forecasting, the apparent scientific exactness conveyed by a one number forecast can conceal a wide range of uncertainty. For example, in the UK a baby is born on its predicted due date just 4 per cent of the time.[4] The most likely lifespan of English and Welsh males born in 2010, according to the Office for National Statistics, is eighty-five years. But the detailed forecast tells us that only 3.9 per cent of them are forecast to have this lifespan. As the economist John Maynard Keynes once said, 'it's better to be roughly right than precisely wrong'.[5]

Sometimes displaying uncertainty can result in a totally un-informative forecast. The chart below is a forecast of the next number to be drawn in the Lotto game run by the National Lottery in the UK. All the numbers have an equal chance of being drawn at any given time (1 out of 59 or 1.69 per cent) so all the bars have a uniform length. Unsurprisingly, this is called a uniform-probability distribution. However, there's no point in being dishonest: if we are truly uncertain about what might happen we need to say so, otherwise important plans may be built on false assumptions. Sometimes such uncertainty is unavoidable. Indeed uncertainty is information in itself. If we know we are facing it, we can try to design robust strategies that will ensure we survive or even prosper whatever the future throws at us.

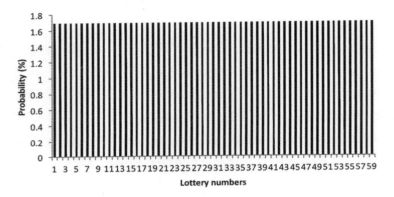

Probability of each number being drawn in the UK lottery

Our innate resistance to the monotonous uncertainty like that displayed in the lottery chart was reflected by the *Daily Mail* when the British lottery started in 1992. It suggested a series of lottery-winning strategies. These included selecting balls with numbers greater than nine. It was argued that these would be more likely to come up because they had two digits written on them and hence more ink. This would make them slightly heavier so they would fall faster and get into the draw more often. Superficially the results up to March 2015 support this.[6] Balls numbered between one and nine were each drawn on average 239.4 times, while those numbered from ten upwards were each drawn 246.6 times on average. But this is easily explained by random variation. If we toss a fair coin twenty times we should not be surprised if we get twelve heads and eight tails. In fact, the 'neat' result of exactly ten heads and ten tails will only happen about 18 per cent of the time. As we have seen, in forecasting we must be careful not to be deceived into seeing systematic patterns in randomness. The chart is the best forecast of lottery numbers, boring though it is.

Another potential problem with forecasts involving probabilities is that people might misunderstand them. In the US, weather forecasters have been broadcasting forecasts with probabilities since 1965 and, as we've seen, their estimates of the chances of rain or sunshine are highly reliable. Despite this, some studies suggest that people misinterpret these forecasts. When pedestrians in five cities (Amsterdam, Athens, Berlin, Milan and New York) were asked to interpret what a forecast meant when it said 'there is a 30 per cent chance of rain tomorrow' only the majority of New York respondents made the correct interpretation: 'there is a 30 per cent chance that it will rain tomorrow at any given location within the specified region'. For example, if you received this forecast for New York, and planned to spend all of tomorrow sitting on a bench in Central Park, there is a three-in-ten chance that you'd at least see a trace of rain during the day and a seven-in-ten chance that you'd stay perfectly dry. It would be akin to tossing a ten-sided die, with three faces showing 'Rain' and seven faces showing 'Fine'.[7] However, some people questioned in the other cities thought that the forecast meant it would rain for just 30 per cent of the day; others believed it would rain in just 30 per cent of the region being covered by the forecast. One person thought that: '30 per cent means that if you look up to the sky and see 100 clouds, then thirty of them are black'.

Misunderstandings of probabilities are not confined to members of the public. Economists were unable to make correct interpretations of the output of economic forecasting models, which involved probability information, in one study.[8] Note that the incorrect interpretations of the weather forecasts are actually expressions of certainty. It will rain for 30 per cent of the day tomorrow. It will rain over 30 per cent of the region tomorrow. Our dislike of uncertainty may, at least in part, be the cause of these misperceptions of the forecasts. The correct interpretation involves uncertainty – we don't know whether or not we will see rain tomorrow if we spend the day at a particular location.

All of this means that when it comes to forecasting, many people prefer just to see a single number – a so-called point forecast – or a single event being predicted with any uncertainty omitted. Obligingly, when forecasting numbers, computers generally produce a point forecast that is equal to the average outcome that would be expected if the period we are forecasting was repeated a large number of times under the same conditions. For example, if we are forecasting next week's ice cream sales, the point forecast would tell us on average how much ice cream we would sell if there were a large number of weeks that had the same weather conditions and the same amount of advertising for our product as next week. When the Bank of England made the growth forecasts that we saw earlier they realised that the factors that would determine growth in quarter three of 2018 could play out in lots of different ways. But if the quarter could be experienced a large number of times, on about 19 per cent of occasions growth would be less than 0 per cent, on about 19 per cent of occasions it would fall between 0 per cent and 1 per cent, and so on. The Bank estimates that the average of all these possible growth levels would be 1.57 per cent. If they wished they could simply use this as their headline forecast, but it would tell us a lot less than their probability distribution.

If we abbreviate forecasts for events that lend themselves to names rather than numbers (e.g. winning presidential candidates, winning horses in a race or next year's bestselling product) we usually have no choice but to state the most likely event. When forecasting the winner of the 2012 US presidential election, taking the average of Barack Obama, Mitt Romney and the five other candidates would not have made sense. We could perhaps have averaged their heights but that would have been an imperfect (though, surprisingly, not a totally useless) guide to who would win.

Only quoting point or single event forecasts, without any acknowledgement of uncertainty, might make forecasts more palatable for forecast users but it has its perils for forecast providers. The Met Office attracted widespread criticism from the media when it forecasted that April 2012 would be 'drier than average' and the month proved to be the wettest April on record. 'Why you shouldn't rely on a Met Office forecast,' advised Olivia Goldhill in the *Daily Telegraph* citing this, and other examples, of apparently 'disastrous' weather forecasts.[9] 'The outlook … well, we're not sure actually,' mocked a headline in *The Independent*.[10] Yet it turned out that the Met Office's internal forecasts were not necessarily wrong. They showed that there was only a 25 per cent probability of very dry weather. The actual event – very wet weather – was judged to have had about a 15 per cent chance of occurring. The forecasters had made the mistake of just publicising the most likely event without any reference to its probability. By simply saying: 'there'll be a drier than average April' they may have given the welcome impression to many people that this was almost bound to occur. The abbreviated forecast may have pleased a few newspaper editors, anxious to keep their stories simple and readable, but these same editors were quick to bite back when the 'barbecue' April failed to materialise.

If forecasters must communicate single numbers or events then it's best if they define exactly what their forecast is: 'we expect that the average number of goals for games in this league will be 2.76' or 'the most probable number of goals is two', or 'the average level of washing machine sales in market conditions like the ones we anticipate next week is 332 units'. Better still, if they are giving the most probable event, they could tell people how probable they think it is: 'two goals is most likely with a 22.6 per cent probability'. Just calling it a forecast is not enough – 'forecast' can be a weasel word. When a weatherperson says: 'I forecast rain tomorrow' people might think they are sure it will rain. Or perhaps they are saying that rain is not certain but it's a lot more likely than no rain. Or perhaps rain is simply the most probable event, though it's only slightly more probable than no rain.

### Less abbreviated forecasts

When numbers are being forecast, there's a less severe way of abbreviating forecasts. Instead of presenting just a single number, forecasters can give a range of numbers that they think has, say, a 90 per cent probability of including the actual outcome. These are called prediction intervals. For example, a 90 per cent prediction interval for tomorrow's temperature at midday in London might be 3°C to 5°C. This means that we think there's a 90 per cent chance the interval will capture the actual temperature in its 'net'.

We can combine these intervals with a point forecast. For example, if our point forecast is 4°C then we are telling people that, to allow for uncertainty, we have a margin of error of plus and minus 1 degree C and there is only a 10 per cent chance that this margin of error won't include the actual temperature.

Prediction intervals at least give us some indication of the forecaster's confidence in their point forecast. For example a 90 per cent prediction interval of -20°C to 20°C would tell us that they are very uncertain about what is going to happen; an interval of 3.5°C to 4.5°C, that they are highly certain. There's some evidence that people make better decisions when they are given prediction intervals rather than point forecasts.[11]

But there are downsides to using prediction intervals. Because they only give partial information they aren't useful in all decisions. For example, the computer tells us that a 90 per cent prediction interval for next week's sales of a product we are selling is 654 to 703 units. We want to make sure we have sufficient stocks of our product at the start of the week so that there is only a 2 per cent risk of running out of it and disappointing our customers. So how much stock should we hold? The 90 per cent prediction interval can't tell us, but a forecast showing a full probability distribution can. We might be able to adjust the width of our interval to suit our purposes (e.g. use a 96 per cent interval in this case), but most forecasting software products only provide intervals for a fixed number of probability values (e.g. 90 per cent, 95 per cent or 99 per cent).

Then there's the problem of getting people to accept prediction intervals. They have a rather featureless profile. We can't see where within the interval the most probable temperatures or sales are, or how far either side of the interval the range of possibilities might stretch. People are likely to regard very wide intervals (like the -20°C to 20°C above) as uninformative and even useless, even if they are an honest statement of the uncertainty being faced. They may even think that the forecaster providing them is incompetent. In one study,[12] people even preferred a narrow prediction interval that did not include the true value to a wide one that did include it. When given two prediction intervals for the number of countries who were members of the United Nations in 1987, 90 per cent of respondents preferred the interval 140 to 150 over the wider interval of 50 to 300, even when told that the correct number was 159. In some circumstances, such as financial forecasting, it seems people will tolerate intervals up to a certain width, acknowledging that the forecasts are subject to unavoidable uncertainty. But their tolerance is limited and, as soon as they get too wide, the credibility of the forecasts disappears.[13]

## Getting real about uncertainty

Point forecasts, or single event forecasts, are like candies. They are easily consumed and they appeal to our reward systems, given our liking for apparent certainties. They give us a short-term fix but leave us open to future shocks; with sweets the surprise might be at the dentist or on the bathroom scales. Point or single event forecasts deny us the opportunity to make rational decisions in the face of uncertainty. The forecast says rain, but should I risk leaving my umbrella at home? The computer says the Chicago White Sox will win today, but should I risk a punt? The sales forecast says we'll sell 200 units, but how many should I produce in case sales are unexpectedly high? In particular, they tell us nothing about the risks of extreme events occurring – those rare events which, when they do happen, can wreak havoc, such as a collapse in the stock market or a flood the height of a house.

Yet point forecasts continue to dominate fields like business

and economic forecasting and (outside the US) weather forecasting – and the work of forecasting researchers in academia.
In a 1987 survey, 90 per cent of US companies said they usually used point forecasts for their sales planning.[14] There's little
evidence that things have changed. Many academic surveys
of company practice have not even bothered to find whether people are expressing uncertainty on their forecasts. One
recent survey of Canadian companies found that 28 per cent
said they used prediction intervals,[15] but that figure looks surprisingly high in my experience. Like the rest of us, managers
don't like forecasts expressing uncertainty. Some have even
been known to query why a point forecast they've received is
not 100 per cent correct in view of the salary they are paying
the forecaster. Even a well-known and acclaimed book, published in 2001 and listing the principles of good forecasting
based on the work of forty world-leading experts, only goes so
far as to recommend the use of interval forecasts and its focus
is predominantly on point forecasts.[16]

So what's to be done to move the world on to the acknowledgement of uncertainty in forecasting? There are three main
hurdles: creating the forecasts, getting people to accept them
and getting people to understand them.

The key problem with creating probability-based forecasts
is that levels of uncertainty often tend to be underestimated.
This means the range of possible values tends to be too narrow.
The computer might tell us the level of a river is bound to be
between three and eight feet when, unknown to anyone, two

to twelve feet is the real possible range. One reason for this is that we have limited past data to work from so very rare extreme events don't get a chance to appear. The once-in-a-thousand-years flood is unlikely to appear in the last hundred years' worth of data. As Chapter 2 demonstrated, people suffer from the same problem so they also tend to estimate ranges of possible outcomes that are too restricted, though there are sometimes different reasons for this.

Even when forecasters have an idea of the possible range, they might estimate probabilities that are too low for extreme events. It's common for forecasters to assume that the 'bell-shaped' forecast shown below is applicable. This is the so-called normal or Gaussian distribution. Like many inventions or discoveries the distribution was not named after the person who discovered it originally. French mathematician Abraham de Moivre formulated the mathematics of the distribution in 1733, but his papers were lost until 1924 by which time it had been named after Carl Friedrich Gauss, who discovered many of its properties. A picture of the distribution even appeared on a German ten Deutschmark bill next to a portrait of Gauss showing how important it has been to scientists and statisticians.

Sometimes there are sound theoretical reasons why the normal distribution is appropriate in certain forecasting situations, but often it is used unquestioningly, merely because it is convenient from a mathematical perspective. Nassim Taleb has even referred to the distribution as the 'Great Intellectual

Fraud'. One problem is that in some circumstances, such as forecasting in financial markets, larger probabilities should be allocated to extreme outcomes to the left and right of the chart. In particular, stock market crashes are much more likely than the curve suggests. Because of this some financial analysts have made their forecasts using curves with fatter 'tails' than the normal curve so that the probabilities of outcomes to the extreme left and right of the distribution are higher. However, estimating how fat these tails should be is challenging, given the scarcity of data on rare extreme events.[17]

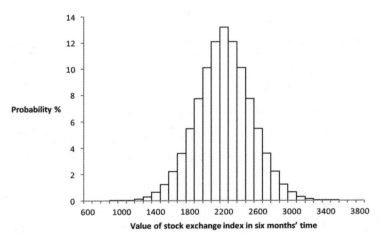

Example of a forecast based on the normal distribution

Robert Merton and Myron Scholes won the 1997 Nobel Prize in Economics for developing a highly sophisticated mathematical formula, the Black-Scholes-Merton model, which could be used to estimate the price of financial options over time (Fischer

Black, who also developed the model, was ineligible for the prize as he had died). The normal distribution lies at the heart of the model and its creators found that it usually provides a useful approximation to price movements. This apparent accuracy encouraged confidence in the model amongst traders and financial institutions and eventually they became oblivious to its key limitation: it grossly underestimates the risk of extreme movements. Taleb[18] has estimated that, if the normal distribution was a correct representation of the relevant probabilities, a stock market crash as large as the one that occurred in 1987 would only take place once in several billion lifetimes of the universe. But people working in the financial sector forgot this and came to regard the formula as a latter-day King Midas. They took huge speculative risks. In 2008 suddenly it all went wrong and the world's financial system was plunged into a catastrophic crisis.[19] Identifying the correct probability distribution for a forecast can be more than a mere academic exercise. Get it wrong and the entire world can be affected for many years to come.

## Making probabilities palatable

Even if we can obtain probability-based forecasts that are reliable, how can we make them more acceptable and understandable? One temptation is to replace numerical probabilities with words that we all use such as 'probable', 'likely', 'very unlikely', 'doubtful', 'small chance' or 'virtually certain'. But words can be vague. Sometimes people might hide behind this vagueness when trying to soften a forecast they know will be unwelcome. A doctor might say 'it's possible that you have a serious disease' when she thinks it's highly probable.[20] Ambiguity like this can also inadvertently cause misunderstandings. Your 'possible' might be my 'close to certain', even if we are agreeing on what the chances are. A reference to 'likely' might be interpreted as a probability of anywhere between 40 per cent and 80 per cent.[21] 'There's a chance' might be somewhere between 25 per cent and 65 per cent. 'Unlikely' or 'improbable' might even be interpreted as 'the event isn't going to happen so we can ignore it'.[22]

However, one study found that the uncertainties involved in forecasts of global warming produced by the Intergovernmental Panel on Climate Change (IPCC) were best conveyed to the public when they included both words and numeric probabilities. For example, people understood phrases that said an event is: 'very unlikely, that is having a probability of 2 per cent to 10 per cent'.[23] Diagrams can also be helpful. Pictograms showing 1,000 outlines of men and women with two people coloured in red to show what a 0.2 per cent risk of a disease looks like are now being used by health professionals to convey uncertainty.

Perhaps the most radical suggestion for conveying uncertainty in forecasts has come from Robin Hogarth of Universitat Pompeu Fabra in Barcelona and his then research student Emre Soyer.[24] They advocate the use of computers enabling people to see simulations of different future outcomes. For example, if there is a forecast of a 30 per cent chance of rain tomorrow, the computer will generate weather outcomes for sets of days that conform to this probability. This might look like: rain, fine weather, rain, fine weather, fine weather, fine weather and so on. That way people actually experience the proportion of days when it rains. As we saw earlier, there is plenty of evidence that we are good at using experience to build up a mental record of the frequency with which different events occur and hence the probability that they will occur on a future occasion. In addition, people's active involvement in witnessing the simulated events seems to lead to a deeper understanding than the passive process of simply being told what the probability of an event is. In a series of experiments it was found that simulated experiences dramatically improved people's ability to estimate probabilities, even if they had no training in statistical methods or probability theory.

These results are promising, but more research is needed to find ways to estimate probabilities, particularly probabilities of rare but high-impact events, and to encourage people to accept and understand forecasts that include probabilities. Sometimes it may not be possible to estimate probabilities because we have no data or awareness of possible future events. We will look at how we can address these situations in Chapter 11. However, if we are prepared to acknowledge and live with uncertainty we will be better able to protect ourselves from its downsides and to exploit the opportunities afforded by its upsides. Forecasts that convey uncertainty honestly and accurately can help us achieve these goals.

# CHAPTER 9

# THOSE WHO
# NEGLECT HISTORY

## Uniqueness everywhere

These days you can place a bet on almost anything. At the time of writing you could bet on Oscar winners, Prince George's first word, when he will first be photographed in a nightclub, the year when alien life will be proved and how Julian Assange, the founder of WikiLeaks, will leave the Ecuadorian embassy in London – the options include a British police car, a laundry bag, a tunnel and a jet pack.[1] You may know the answer to some of these cliffhangers by the time this book is published. Of course, most bets relate to sporting events. At least £10 billion worth of bets are placed on horse racing each year in the UK,[2] while betting on football, both legal and illegal, has grown into a billion dollar global industry.[3] Even when we don't formally place a bet, each time we buy a new house or used car we are implicitly taking a gamble and also, at least implicitly, making a forecast of how things will turn out.

All these situations may appear to be unique. They all have their own combination of characteristics that make them different from other situations. Hundreds of horse races take place each year – but each involves a different field of horses and different conditions. Julian Assange's escape, in particular, looks as if it might be unique. We don't have 100 occasions where he has already left the Ecuadorian embassy, let alone 15 per cent of which were via a laundry bag, 10 per cent through a tunnel, and so on. This suggests that we can't analyse past data looking for patterns that might give us some idea of what may happen.

However, uniqueness is in the eye of the beholder. Other people have sought asylum in embassies around the world and eventually managed to leave. The number of people may be too small and the number of situations too varied to permit us to carry out a reliable statistical analysis, but they may provide some guidance if we have to make a forecast based on judgement. Alternatively, we may be able to view the situation from a different perspective. We could look at a wider range of situations where people have been fugitives from the law to see what percentage eventually gave themselves up. Or we could examine the number of times the legal authorities have eventually dropped charges.

Innovative new products, when they are launched, may seem unique but their future sales patterns have similarities to those of earlier innovations. When televisions first appeared their adoption patterns had similarities to those of radios when they were first launched. New pharmaceutical products have sales patterns similar to drugs that were launched in the past. Each new American presidential election may look different, but every election since 1860 has followed a common pattern which means it's been possible for some forecasters to repeatedly forecast the winner with perfect accuracy.[4] Even conflicts, such as wars and strikes, have analogies in the past and, if we analyse their underlying characteristics, we can often make reasonably accurate forecasts of their outcomes.[5]

Despite this, we have a tendency to see each new situation as being a one-off.[6] A marketing manager might be convinced that next week's sales promotion will be uniquely successful. 'Forget past campaigns. This one will be a scorcher!' he might declare. But cool reflection might indicate that earlier campaigns contain lots of pointers to its likely success. Henry Ford famously said: 'History is more or less bunk.' When we forecast we often adopt the same belief. As a result we downplay information contained in past data which would give us guidance on the probability of different outcomes occurring in the future. Sometimes we ignore the past completely in favour of intuitive judgement.

Ignoring the statistics of past cases (or the statistical 'baserate') can make us over-optimistic about the future. Marriage seems to be a particularly optimistic institution if we consider the base rates. In the UK, 33 per cent of marriages in the years

1995 to 2010 ended in divorce.[7] In the USA the divorce rate for third marriages is between 70 per cent and 73 per cent.[8] Yet, presumably most people getting married will think that these statistics don't apply to them because the uniqueness of their relationship makes them different.

When setting up a new business, entrepreneurs tend to focus on what they perceive to be their distinctive qualities. A new restaurant owner might reflect on what's special about the meals that will be offered on his menu, the skills of the new chef he's just appointed, the effort he has put into the design of the restaurant and the encouraging comments of his new customers. He might rate his probability of still being in business in a year's time as 95 per cent. However, statistics might reveal that only 56 per cent of restaurants in his town survive for at least a year – a base rate that should at least carry some weight in his forecast.

In one study,[9] people who were in the process of setting up new business ventures were asked to estimate the probability that their enterprise would eventually be operational. Their average estimate was 90 per cent. Yet only 47.8 per cent of the prospective ventures actually became going concerns. For the businesses that did become operational, first-year sales were overestimated by an average of 62 per cent. Of course, it's natural for people engaged in setting up a new business to be optimistic. After all, why go to the trouble if we believe we are going to fail? However, the interesting finding emerging from this study was that it was the preparation of projected financial statements that led to the optimism bias. Rather than making their forecasts more realistic, this caused the intended entrepreneurs to focus on the specifics and details of their particular venture and how any potential obstacles to success might be overcome. This distracted them from taking what psychologists have called 'the outside view' where they would see their business venture in the context of many similar ventures, a significant percentage of which had failed to take off.

Optimism bias is particularly prevalent in the construction industry. The Scottish Parliament at Holyrood had an original forecasted construction cost of £40 million – the final bill was £414 million. This incredible hike in costs put it ahead of projects like the Channel Tunnel, the British Library, the Humber Bridge and London's Jubilee Line extension in the infamous league table of projects with massive cost overruns. However, even the Scottish Parliament couldn't compete with the Sydney Opera House, Concorde and the Barbican in London.[10] The Danish economic geographer, Bent Flyvbjerg, reported in one study that inaccurate forecasts of costs for transportation infrastructure projects averaged 44.7 per cent for rail, 33.8 per cent for bridges and tunnels and 20.4 per cent for roads.[11] Flyvbjerg acknowledges that some of this under-forecasting may be due to what he kindly calls 'strategic misrepresentation' where costs are deliberately underestimated by contractors to secure projects. But much of it, he argues, is due to planners seeing each project as unique. 'The thought of going out and gathering simple statistics about related projects seldom enters a planner's mind,' he says. 'Planners may consider building a subway and building an opera house to be completely different undertakings with little to gain from each other. In fact the two may be – and often are – quite similar in statistical terms, for example as regards the size of cost overruns.'

As a cure for this expensive problem Flyvbjerg recommends the relatively simple technique of reference class forecasting. To apply this, a forecaster first identifies a collection of similar projects to the one that is in prospect. This is called the reference class. This collection must be large enough to contain a wide variation in outcomes, for example variations in construction costs, but not so large that it contains projects that are vastly different from the anticipated project. For example, a reference class for a new metro light railway might contain details of previous projects involving the construction of high-speed and conventional railway lines, together with other metro light railway projects and schemes where guided buses follow tracks. An analysis of the projects in the reference class might then suggest a forecast like the one below.

| Cost Overrun | per cent of Projects |
|---|---|
| 0 to under 20 per cent | 45 |
| 20 to under 40 per cent | 25 |
| 40 to under 60 per cent | 20 |
| 60 to under 80 per cent | 10 |

When a local government authority receives the estimate of construction costs from a railway contractor it can now gauge the risk that the project's cost will overrun by different amounts. For example, there is a 45 per cent chance that the overrun will be less than 20 per cent, but a 10 per cent risk it will be between 60 per cent and 80 per cent. If the contractor estimates a construction cost of £100 million, and the government only wants to accept a 10 per cent risk of having to pay more than it budgeted for, it can revise its budget for the project upwards by 60 per cent to £160 million.

## No end of trends

Sometimes a novel situation can arise as a result of gradual changes. A public health campaign may not have an instantly noticeable effect. But each month more and more people take notice of the campaign and so in the following months we will see fewer people engaging in the unhealthy behaviour that the campaign is trying to eliminate. Each year, because of road congestion and environmental concerns, people may be taking more journeys on trains. In both these cases the future may be different from the past. Next year, fewer people may be smoking than ever. Next month's number of passengers may be unprecedented. Nevertheless, we can still learn from the past by studying the momentum with which things have been changing and extrapolating this to obtain our forecasts.

However, just as we are too quick to spot uniqueness we can also be too quick to see trends. Two data points don't make a trend but we may be tempted to see one, nevertheless. This temptation is likely to be particularly strong if we are rewarded for making improvements. Newspaper stories such as 'Devon and Cornwall Police report 5 per cent drop in crime last year'[12] and 'rail punctuality improves: the percentage of trains running on time in autumn 2004 showed a 3.9 per cent improvement on previous year'[13] look pleasing. But crime rates and rail punctuality year-to-year are bound to vary partly because of random factors like the weather. It's a bit like throwing a six followed by a four on a die and assuming there's a downward trend in the scores. As Nobel Laureate Daniel Kahneman has said: 'We form powerful intuitions about trends ... on the basis of information that is truly inadequate'.[14] We extrapolate these apparent trends into the future at our peril.

Even when we see more than two data points there's still a danger that we'll see illusory trends. Look at these numbers and see if you can forecast the next one in the sequence:

1840, 1860, 1880, 1900, 1920, 1940, 1960...

This looks like it might be a typical question from an IQ test and if it was then 1980 would, of course, be the answer because there's a pattern in the numbers – they increase by twenty. But this is a sequence of numbers representing real events: every US President who was elected in each of these years had the misfortune to die in office. These included W. Harrison (1840), Lincoln (1860), Garfield (1880), McKinley (1900), Harding (1920), F. D. Roosevelt (1940) and Kennedy (1960). Some people had an astrological explanation for this pattern – conjunctions of Jupiter and Saturn roughly correspond with these dates. Others believed that Indian chieftain Tecumseh or his brother Tenskwatawa, had cursed the presidential office in revenge for a military defeat in 1811.

However, Ronald Reagan was elected in 1980 and, despite an assassination attempt in 1981, he did not die in office nor did George W. Bush, who won in 2000. There was, of course, no underlying reason for the sequence; it was purely coincidental. But as we found in Chapter 2, our brains are adept at searching for patterns and then providing a ready explanation for their occurrence.

## Trends without end

Even if a trend is genuine, once we've experienced it for a long period of time we sometimes develop an unquestioning belief that it will continue. Generations of people who have never experienced anything other than improving standards of living may see this as the norm and never imagine things could change. Most people in the UK expect to live longer than their parents because life expectancy has been increasing since the nineteenth century. For the past twenty-five years or so most Western economies have experienced relatively low levels of inflation and it's tempting to think the bad old days of the 1970s, when inflation rates exceeded 20 per cent, will never return.

But trends that seem to be a permanent fixture do change. Before the financial crisis of 2007–08 the Standard & Poor's credit rating agency used a housing market model that would not accept a negative number because it assumed house prices would never decline.[15] Yet after peaking in 2006, the US housing market saw the biggest crash in ninety years – typical prices were down by 15 per cent by the end of 2007 according to one index.[16] Similarly, the period 1948 to 1973 saw real crude oil prices decline. Governments and companies based their strategies on the assumption that this trend would continue. Electricity companies switched from burning coal to burning oil. But in 1973 it all changed. In response to American involvement in the Yom Kippur War, Arab countries imposed an oil embargo. The price of oil quadrupled and a new upward trend began which lasted until 1981.

Nevertheless, sometimes there's a good reason to believe, cautiously, that past sequences or trends will continue into the future – at least into the short-term future. This is the case when there are forces which on balance are pushing the graph in a particular direction over time and we can reasonably expect them to continue. These forces might be factors like technological advancement, human population growth, improvements through learning or cultural changes. Consider the following graph which shows the annual casualty rate on UK roads for the period 1984 to 2011. Factors such as safer cars, better designed roads and road safety campaigns have all contributed to forcing the graph downwards despite a doubling in the annual number of vehicle miles travelled – a remarkable achievement. If these forces continue to operate then we can expect the trend to continue, though the rate of decline is likely to level off as it gets closer to zero.

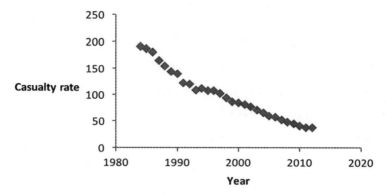

No. of people killed or seriously injured
per billion vehicle miles 1984–2012[17]

To make a forecast of the casualty rate for a future year we could simply extend the trend line downwards, but given the apparent slowing of the downward trend in recent years this might produce a forecast that is too low. Extend the line too far and we'll be forecasting a negative casualty rate. Indeed, the rate of decline seems to have varied slightly over the years so it seems too simplistic to represent the decline as a 'straight line' over the whole period. Also, we know that random factors will affect the future casualty rate. For example, a severe winter might cause more accidents as vehicles run into snow or skid on ice; an exceptionally mild one might lead to fewer accidents than normal. We therefore need to take into account the uncertainty caused by these factors as well.

Modern computer-based forecasting methods can detect changes in trends. They do this by adjusting their current estimate of the trend as each new figure arrives. If this figure is higher than expected then they will increase their estimate of the trend and vice versa. These methods can also estimate the future uncertainty we will face and they assume this uncertainty will increase as we forecast further into the future. Here's a forecast for 2015, based on the data up to 2011.[18]

| Forecast casualty rate in 2015 | Probability per cent |
| --- | --- |
| under 3 | 0.9 |
| 3 to under 9 | 3.8 |
| 9 to under 15 | 11.1 |
| 15 to under 21 | 21.1 |
| 21 to under 27 | 26.2 |
| 27 to under 33 | 21.1 |
| 33 to under 39 | 11.1 |
| 39 to under 45 | 3.8 |
| 45 and over | 0.9 |
| Total | 100 |

Forecast of the casualty rate on UK roads for 2015

By 2011 the computer figures that on average the casualty rate is falling by 4.5 deaths and injuries per year for every billion miles travelled. Using this, it forecasts that the most likely casualty rate in 2015 will be twenty-four. However, it recognises that random factors may lead to a different outcome and the above table shows estimates of the probabilities that the rate will fall within different ranges.

So what is the forecast claiming? Well, all it's saying is that, if past trends continue and the variation caused by random factors is the same as we have observed in the past, then these are the probabilities of different outcomes occurring – nothing more and nothing less. The forecast is not claiming any magical or mystical powers to see into the future. Only fortune-tellers and their like would claim that. We have the right to expect that the forecaster has used the most accurate and up-to-date information available at the time of the forecast, as long as the cost of getting this information can be justified. We can expect that they have used the most appropriate method to distil the recent underlying trend and to estimate the probable effects of random factors. We can expect them to be honest about what assumptions they have made. We can't reasonably expect any more.

### We don't even know what's happening now so how can we forecast the future?

Forecasting the future is challenging enough, but there are some circumstances where we don't even know what's happening now. When forecasting change we need to know where we are currently at – after all, what are we changing from? In some circumstances it takes a while to obtain data on the present situation and by the time we have the accurate numbers things have moved on. Estimating what's happening at the present time is called nowcasting and this is particularly problematic for economic forecasters. Provisional figures about the state of the economy are frequently revised as more data is collected and checked. One study[19] found that in the USA initial estimates of growth in gross domestic product (GDP) had later on been revised by up to 7.5 percentage points in the years from 1961 to 1996. There is evidence that errors like these in the initial estimates of the current state of an economy can have a considerable negative effect on the reliability of economic forecasts.[20]

This may not seem an issue for weather forecasters. After all they can easily measure today's temperature, wind speed or rainfall. The problem is that these measurements can never be exact. The weather is known as a chaotic system, which means that even very slight changes in current conditions can make a big difference to tomorrow's weather. The well-known butterfly effect, where a butterfly flapping its wings in one continent can allegedly cause a hurricane two continents away, exemplifies this. As a result, tiny errors in today's atmospheric measurements can lead to large errors in forecasts made for a few days ahead.

To cope with this, weather forecasters use what are known as ensemble forecasts. They begin by making an estimate of what the true starting conditions are and use their weather model to predict what will happen for the day in question. Then they tweak the starting conditions to reflect the level of uncertainty that they have about the current conditions and, for each new set of conditions, they produce an alternative forecast. If all the forecasts they produce look similar then this suggests there is little uncertainty associated with the forecasts they are making. If the forecasts vary widely then the forecasters will be less confident in predicting what will happen. This method allows forecasters to attach probabilities to their forecasts. If they make twenty-four forecasts, each based on a different set of current conditions, and eight of these predict more than 3mm of rain in a particular location during a particular time period, then the forecasters would estimate there is an eight out of twenty-four (i.e. 33 per cent) chance of this event occurring.

**Self-destructive forecasts: when forecasts cause change**

Sometimes the future can differ from the past for one simple reason: the act of forecasting the future itself causes things to change. In 2001 the British Labour Party, with Tony Blair as Prime Minister, had been in office for one term. Blair decided it was a good time to call another general election. It was widely expected that Labour would secure a substantial share of the vote and easily secure a second term. Throughout the election Labour maintained a substantial lead in the opinion polls. On the eve of the election one poll suggested the party would win 47 per cent of the vote, enough for Blair to continue in government, given the quirks of the British electoral system. When I fitted a model to patterns observed in past elections this also suggested that a vote share of 45 per cent to 48 per cent was the most likely outcome.[21]

So how did Blair do? His party got only 40.7 per cent of the vote. It's unlikely that a forecaster would be celebrating this result. Although the media declared Blair's victory a landslide, his share of the vote was much lower than expected and it fell in a range that my computer model said had only a 5 per cent chance of containing the actual result.

So was the forecast wrong? As we'll see in Chapter 10 we can seldom judge the quality of a forecast on one result. My forecast didn't say that this share of the vote was not going to occur; it just said it was unlikely. But there is an important problem here: the curse of the self-destructive forecast.

When political analysts pored over the voting data they realised that four out of ten people had not bothered to vote – the lowest turnout since 1918. Because a strong Labour victory had long been predicted in the polls and by media commentators, many people thought that voting would not make any difference to the outcome so they stayed at home.[22] This was particularly the case amongst those who would have voted for Labour.[23] So the act of making a forecast changed people's behaviour and the forecast ended up appearing less reliable.

This can be a common problem. An analysis of past data might suggest that a company has a 90 per cent probability of going bankrupt in the next twelve months. But this prediction is likely to galvanise a team of sleepy managers into action and spur them into doing all they can to prevent the collapse. A year later, we find that the company has survived. Forecast that an area of town might be the new crime hotspot and the authorities might take measures, like increased policing, to prevent this happening.

In contrast, some forecasts are self-fulfilling. In Greek mythology Laius, the father of Oedipus, was aware of a prophecy that he would be killed by his son so he left the baby boy to die on a mountainside. The boy was rescued by shepherds and grew up not knowing who his father was. Years later he had an argument with a stranger, who was riding in a chariot, about who should give way on a narrow road. The quarrel led to Oedipus killing the stranger. Oedipus later discovered that the man in the chariot was his father. The prophecy had become self-fulfilling.

If a stock market guru forecasts that a stock's price will rise, people will buy it and so the price will probably rise, at least in the short term. Two *Daily Mirror* financial journalists were convicted in 2005 for buying shares and then tipping them in their 'City Slickers' column twenty-four hours later. Investigators were alerted when the journalists bought thousands of pounds worth of shares in Sir Alan Sugar's Viglen company, before endorsing Viglen the following day in a column headed: 'Sugar to Head Next Gold Rush.'[24]

So no matter how sophisticated a forecasting method is it can still produce inaccurate predictions, as our behaviour may be influenced by the forecast, leaving the poor forecaster with egg on their face. Forecasts therefore tend to be more reliable when the thing we are forecasting behaves independently of the forecast. Halley's Comet won't change its course because we've forecast that it will next appear in 2061, and presumably most customers won't change their purchases because a company has internal forecasts of who will buy its latest product.

Interestingly, the interaction between behaviour and forecasts can work both ways – our behaviour can cause havoc with forecasts. Launched in 2008, the Google Flu Trends web service was designed to make predictions of influenza activity in over twenty countries. It bases its predictions on the number of searches made on Google using terms like 'medicine for fever' and 'cough'. There is evidence that these predictions can in general be highly accurate. However, in 2013 publicity about the service led many people to enter flu-related search terms, not because they were ill, but because they wanted to find out more about the prediction methods used. This led to Google Flu Trends forecasting a huge rise in the number of influenza cases and grossly overestimating the actual number of people who subsequently suffered from the illness.[25]

### Is there anything new under the sun?

The late Ralph Harris, Lord Harris of High Cross and former head of Britain's Institute of Economic Affairs, held a number of controversial views. An ardent admirer of Margaret Thatcher, a Eurosceptic and a campaigner against smoking bans on trains, he was also a forecasting sceptic. He once wrote: 'All forecasting is in an important sense backward-looking

– vividly compared to steering a ship by its wake'.[26] To some people, the time statistical forecasters, in particular, spend in carefully analysing past trends and relationships suggests they have more in common with historians than people who focus on the future. Clearly, there are problems when we become obsessed with past data. Rare events may not show up on our records so we discount the possibility of them occurring in the future. And, through the lens of the past we may not foresee the emergence of novel events, such as the development of the internet. Both statistical methods and humans are constrained by the past and present when they try to look into the future. Humans are particularly influenced by the very recent past and present. In workshops where managers were asked to write scenarios of how the future might unfold, current concerns that had been highlighted in the media, such as terrorist activities, tended to predominate.[27]

In his critique of historicism, the notion that there are rhythms, patterns, trends and laws that 'underlie the evolution of history', philosopher Sir Karl Popper argued that the course of history is closely associated with the growth in human knowledge. But no society can anticipate today what it will only know in the future. Hence, it is impossible for us accurately to predict the future course of human history.[28]

But not all forecasting is about what will be played out on the huge canvas of human history. It's about tomorrow's weather, next week's sales of a product, who will win today's soccer match, our own personal futures and even larger-scale events, such as the risk of terrorist attacks, economic slumps or the future solvency of large corporations. Often the past data we have, or our past experience, does contain useful information to help us to forecast these events, but we need to interrogate the data appropriately and imaginatively to tease out its messages. We need to resist the temptation to discard all but the most recent data or to assume, purely for convenience, that past data fits into neat idealised models, such as the normal distribution.

Registering odd inconsistencies in our current information, or things that don't make sense, can lead us to signs that a surprise may be on the way. In his book, *No One Would Listen*,[29] Harry Markopolos describes how the numbers in Bernie Madoff's financial reports did not make sense. A graph showing returns rising at a 45° angle could not be real. The shocking, and as it happened true, possibility that Madoff was running a Ponzi scheme had to be entertained. The data was there; it just needed someone with insight and imagination to spot the possibility. Harvard Business School professor Max Bazerman and consultant Michael Watkins have shown that shock events such as 9/11, Hurricane Katrina and the collapse of the US energy company Enron could have all been predicted with the information we had.[30] Although the specifics of the events were not necessarily predictable in every case it was possible to foresee the events in a general sense, such as 'terrorist attacks on US airliners' so that steps could be taken to reduce their possibility. Extrapolating current trends and analysing the implications of their interaction can alert us to the possibility that a situation we have never experienced before might be waiting in the future. For example, how might trends such as ageing populations in the West, improvements in health care, increased global energy use, the rise of China as a global economic power, advances in information technology and rising world temperatures jointly play out?

Experience, data and knowledge from the past and present can lead to trustworthy forecasts as long as we acknowledge their limitations and are honest and open about the uncertainties involved. Even the rise of the internet wasn't completely unforeseen. Daniel Bell was a sociology professor at Harvard University and has been described as one of the leading American intellectuals of his time. In 1979 a chapter he wrote in an edited book contained the following paragraph:

> 'The really major social change of the next two decades will come in the third major infrastructure, as the merging technologies of telephone, computer, facsimile, cable television and video discs lead to a vast reorganisation in the modes of communication between persons; the transmission of data; the reduction if not elimination of paper in transactions and exchanges; new modes of transmitting news, entertainment and knowledge...'[31]

Prescience indeed.

# CHAPTER 10

# YOU CAN'T TELL ME MY FORECAST WAS WRONG

## The sceptical boss

Imagine this. One day the boss of a textile company tells me he's going to manufacture clothes for a new market: adult males in the US. One thing he needs to know is the average height of American men over the age of twenty so he asks me if I can come up with a figure. 'I don't want any of your probability gobbledygook,' he grunts, 'just a simple average'. After some Googling I find what looks like a reliable number from a reliable source: 69.3 inches.[1] It's based on a carefully designed representative survey that involved health technicians measuring the heights of 5,647 American males aged twenty or over.

Within a few seconds I give the boss the average, thankful that the internet allows you to retrieve information so quickly.

But he frowns and looks doubtful. 'I've got a friend in the US who is aged over twenty,' he says, 'I'm going to email him and ask what his height is.'

A day later he's on the phone to me, a hint of triumph in his voice. 'Your figure was wrong. My friend is 62 inches tall so you were nearly 12 per cent out. So much for these so-called carefully designed surveys!'

'But what I gave you was an average...' I protest. But he won't have it. 'My prediction: the end of surveys,' he says in an odd echo of Dominic Lawson's diatribe in the *Sunday Times*, as we saw in the Introduction.

Sounds crazy? Well it's no different from how computer-generated forecasts are often judged. As we saw earlier when forecasts are abbreviated to point forecasts, we are usually given a prediction of what will happen on average in the circumstances we think will prevail during the period we are forecasting. For example, suppose that a company has five advertisements for a product on television next week. The computer might say that, in weeks when it has this level of advertising, it will sell on average 30,000 units of a product. When the actual sales turn out to be 32,000 units we cannot say the forecast was wrong. Judging a forecast of an average by how close it is to an individual value makes as much sense as the boss's dabbling with statistics.

The problem is that, in any one period, we are never sure of the true average. Measuring the height of one American male aged over twenty won't reveal the average height of all

American males over twenty. Similarly, the sales figure we see from a particular week does not reveal what average sales are under the conditions that applied in that week. We make a forecast of this average, but we never actually see it so we can't say our forecast was wrong.

Here's another example: recall that an entry in the British National Lottery involves selecting six numbers from the fifty-nine available.[2] If you select 1, 11, 23, 34, 38 and 56, for example, your chances of winning the jackpot are one in 45,057,474 or 0.0000022 per cent. I therefore forecast that you will not win – meaning that if you select these numbers not winning is the likeliest outcome. Amazingly, you win. Was my forecast wrong? Not at all; I simply told you what was most likely to happen and I was absolutely correct.

Unlike the lottery, in most circumstances we don't know which outcome would have been the most probable; we only know the result of the event that occurred. We may blame the horseracing tipster who lost us £100 when we followed his advice and backed the horse he suggested. But, in truth, the horse he predicted would win may have been the most likely winner. If the race could be rerun a hundred times then his chosen horse might have won 80 per cent of the time. We can only say for sure that a prediction was wrong, if a forecaster insisted that an event was certain to occur, rather than just likely, but the event subsequently failed to materialise.

Even though we can seldom tell if a single forecast is inaccurate, woe betide the finance minister who forecasts that the economy (as measured by GDP) will grow by 2.3 per cent in the following year when growth turns out to be only 1.2 per cent. And it's likely goodbye to the weather app on your phone if it forecasts there'll be no rain tomorrow and we find ourselves soaked in a sudden downpour. In addition to annoyance, I suspect that many of us experience a touch of *schadenfreude* when these things happen; it appears that the boffins, despite their much vaunted expertise, are fallible after all.

However, choosing not to trust a forecaster based on one poor forecast would be like sacking a football manager after just one bad match. A fairer way to assess the reliability of a forecaster, or of the method they have employed, would be to judge how well he or she predicts lots of numbers or events they haven't seen (this is called a hold-out sample). For example, if a forecaster has twenty-five years of past data they might take data from the first fifteen years to create their forecasting formula, but omit outcomes from the last ten years in order to test if the formula can accurately predict the values. Despite this, forecasters often recommend their pet methods by reporting their accuracy in predicting pitifully few outcomes. Papers in leading international journals extol the virtues of the researcher's newly invented forecasting method because its forecasts of the average were closer to just one, two or three future individual values than an existing method.[3]

Often, the more complicated the proposed method, the fewer future values it's compared with. It's as if the researchers have exhausted themselves in devising the method and don't have the energy left to compare its forecasts against a large number of unseen or future values. I have even seen a method designed to forecast electricity consumption in China that did not compare its forecasts with actual consumption across a whole year. Only nine months of 2010 were used in the testing. Data for October to December, when consumption is likely to be relatively high, was excluded. Yet the researchers concluded 'our proposed model is an effective forecasting technique for seasonal time series...'[4] These complex methods often have very long impressive names such as 'flexible meta-heuristic framework based on an artificial neural network multilayer perceptron' or 'polynomial curve and moving average combination projection model'. The first of these was tested on eight unseen observations and the second on just two. As I wrote in the journal *Foresight*, this suggests a rule: if the name of a method contains more words than the number of observations that were used to test it, then it's wise to put any decision to use the forecasts from the method on hold!

Even worse are cases where a method's forecasts are not even tested against unseen observations.[5] Instead, a set of observations is used to obtain the forecasting formula and the same observations are then used to assess forecasting accuracy. Not surprisingly, the apparent accuracy of the method is almost always exaggerated – and when it's put to the task of making real forecasts its performance is usually much less impressive. In fact, if we make the forecasting formula complex enough we can make it provide us with 'forecasts' that perfectly match the observations it was fitted to. Suppose the number of babies born in a hospital over four successive weeks is 20, 45, 62 and 38 and we decide to use these numbers to derive a forecasting formula that will predict how many babies will be born in future weeks. The following formula will 'predict' the number of babies born in each of the four weeks perfectly:

No. of babies = $-5.5t^3+29t^2-23.5t+20$ where t is the week number.

For example, in week two it would 'predict': $-5.5(2)^3+29(2)^2-23.5(2)+20 = 45$ births

But we'd be misguided if we concluded that our formula was bound to give perfectly accurate forecasts for the future. Our formula is largely representing the idiosyncratic pattern of randomness that occurred in these four observations (earlier we called this phenomenon overfitting). This pattern will be very unlikely to be repeated in subsequent observations so when we use the formula for later periods we are likely to get extremely inaccurate forecasts of future averages. In fact, if we use it to predict the number of babies born in week five it gives us a forecast of minus 60 births, an event that would be new to science. For week ten the forecast is minus 2,815 births.

Good forecasting formulae filter out the superfluous information so they model only the pattern of underlying averages. It's obvious that four observations are not usually enough to obtain a reliable forecasting formula, but the same principle applies when we use more observations. For example, if we have thirty past observations we'll still be able to find a formula, albeit a more complex one, that fits them perfectly. The famous physicist Niels Bohr is alleged to have said: 'Prediction is very difficult, especially about the future.' Just because we can find a formula that fits our past experience well, or even perfectly, we shouldn't necessarily expect it to perform as well in the future.

### The fake weather forecaster

On 5 May 2015, two days before the British general election, the Betfair site predicted that there was an 83 per cent chance that the Conservative Party would win most seats in the new Parliament. When the votes were counted a few days later, it turned out that the Conservatives had indeed won a majority of seats (in fact even more than most people expected in their wildest dreams). Given Betfair's high probability for an event that actually occurred, their forecast seems a good one. But how can we measure the accuracy of forecasts like this?

Calibration is one way to assess forecasts. Suppose we take 100 days where a weather forecaster had predicted 'a 20 per cent chance of rain'. If it rained on exactly twenty of these days then the forecaster is said to be perfectly calibrated. Similarly, when a forecaster has predicted a 30 per cent chance of rain, for perfect calibration it must then rain on 30 per cent of days, on half the days when a 50 per cent chance has been predicted, and so on. Earlier we saw that US weather forecasters are experts who produce very reliable forecasts. This can be shown by their excellent calibration – though it's not absolutely perfect.

For example, in one study it rained on about 30 per cent of days when they had forecast a 30 per cent chance of rain.[6]

Of course, unlike weather forecasts, we don't have elections every day so there usually won't be enough outcomes for us to use calibration to assess how good Betfair's election forecasts were. If we had probability forecasts for individual constituencies it would be useful (e.g. we might have said in constituencies where we forecast a 30 per cent chance of a Conservative winner, 33 per cent of the time the Conservative won).

But where we do have enough data, there's another potential problem with calibration. Suppose I've got a job as a television weather forecaster for Manchester, England and I'm partly paid based on how well calibrated I am. I secretly got the job under false pretences, having faked my qualifications because I was short of cash. In truth, I know nothing about meteorology but I do know how to cheat. I soon find out from the internet that it typically rains in Manchester 140 days per year[7] – that's 38 per cent of days. All I now need do is turn up each day in the studio and make the same forecast: 'Hi folks, there's a 38 per cent chance of rain tomorrow.' I do this irrespective of whether it's summer or winter and despite other forecasters predicting droughts or monsoons. People soon cotton on. They realise that they can forecast my forecast – he'll be predicting a 38 per cent chance of rain. Eventually, I earn the name 'Mr 38 Per Cent', but I'm happy because I know that in the long run it will rain on close to 38 per cent of days so I'll be well calibrated and my pay cheque will reflect this.

Because of this weakness, meteorological offices use scoring rules to measure the inaccuracy of probability forecasts. The Brier score is one well-known example of such a rule. It's easily calculated for a single forecast as the following formulae show:

If the event occurred: Score = (1-probability forecast)$^2$.

If the event did not occur: Score = (0-probability forecast)$^2$.

The best score is 0 and the worst is 1, so the lower the score the better. For Betfair's forecast that the Conservatives would win the largest number of seats the formula is as follows:

Score = $(1-0.83)^2 = 0.03$

This is nearly as good as it gets. Given that the event in question occurred, their score would have been even better if their probability had been higher – a probability of one would have led to a perfect zero score while a probability of 0 would have led to the worst possible score. However, as with all forecasts, we need to monitor their performance over a period of time before we can have some confidence in their accuracy.

If the Brier score was introduced by the TV company it would be time for me to say goodbye to my Manchester weather forecasting job. The Brier score rewards us when we increase our probability if we think the event is likely to occur and lower our probability when we think the event is unlikely. Keeping to the same probability through storms and droughts would not earn me much money.

## The principled way to judge forecasts
Despite all the problems of measuring accuracy, it goes without

saying that we want forecasts to be accurate. Unless a forecaster has an ulterior motive there's no point in deliberately producing inaccurate forecasts. In some circumstances, when there's lots of data available, we can get a good idea of how accurate forecasts are. For example, electricity companies have long records of the demand for electricity for every half-hour of every day so they can have reliable measures of accuracy.[8] And although the demand will depend on factors such as the day of the week, the time of day, the season and the weather, the underlying patterns will remain fairly stable. This means that accuracy in the past is likely to be a good indicator of accuracy in the short- to mid-term future. If we can measure accuracy we can find which prediction methods lead to the greatest accuracy and we can learn how to improve these methods.

However, the tendency of some researchers to test their forecasting methods on a few unseen outcomes, or even none, is symptomatic of a practical problem that occurs in many other circumstances. Often not enough unseen outcomes are available for us to test the forecasts. If we require twenty years of future data to determine if a forecasting method is accurate, then by the time the forecaster is celebrating their own accuracy or regretting that they ever made the forecasts, it'll probably be too late for anyone else to care. Even when we have some past data available to test a method, what it tells us may by now be out of date, because some fundamental factors have changed. As they say in investment advertisements: 'past performance is not necessarily indicative of future results'.

So is there another way, or at least a complementary way, besides accuracy, of assessing the quality of forecasts? Suppose a man decides to gamble his house on a 200/1 outsider in a horse race. Amazingly, the horse wins and the man enjoys the rest of his life as a millionaire. Did he make a good decision? Many people would answer yes. Others would say the man was extremely rash and only got away with it because he was incredibly lucky. 'He must have been mad!' we can imagine them saying. Rather than judging the decision based on its outcome, those in the second camp are making an assessment of the man's decision-making process. Had he really thought through the risk of losing and the huge consequences for him and his family if this happened?

We can do the same for forecasting: rather than focusing exclusively on outcomes where luck can make an ill-judged forecast look good, we can assess the quality of the process that led to a forecast. Did it make best use of available information? Did it make assumptions that most people would have agreed were sensible at the time? Did these assumptions survive challenges from people arguing that they were wrong? Was uncertainty acknowledged?

The CIA has a similar list of criteria that its analysts should meet when they make forecasts for events like the future state of the world economy, the possibility of terrorist attacks, world politics and scientific developments.[9] In the year 2000, as part of a briefing for the newly inaugurated George W. Bush administration, the CIA put together a seventy-page dossier of predictions as to what the world would be like in 2015.[10] Although Britain's *Daily Mail*[11] reported that they'd not predicted events like the financial crisis of 2007–08, only fortune-tellers, astrologers and their like claim to know exactly what will happen in the future. The rest of us don't possess this mystical knowledge. All we can do is have a forecasting *process* that makes the best use of what we know now. That way we are more likely to be accurate and can't be blamed when the unforeseeable happens in the future.

How can we tell if a forecasting process is a good one? We can't simply rely on what appears at first sight to be sensible. We need evidence. So, we have to rely on researchers who compare different methods and processes in a wide variety of situations. This research should reveal the conditions that are likely to favour one approach rather than another. Sometimes, the results are surprising and occasionally they're unwelcome. In the 1970s, Spyros Makridakis, then a professor at the INSEAD Business School in Paris, compared the accuracy of a range of forecasting methods on 101 business and economic data series. Common sense might suggest that the most sophisticated methods – those able to tease out complex patterns from large amounts of available information – would perform best. Instead it was the simpler methods that were the winners. But when Makridakis initially tried to publish his findings in an academic journal, he was rebuffed. His results could not be correct, it was argued. Clearly, the sophisticated methods had not been applied correctly – they had to be more accurate.

Undeterred, Makridakis gathered an international team of experts and asked them to repeat his experiment, this time on 1,001 series. The main findings were the same – simple methods could be as accurate as advanced methods.[12]

More recent work by Makridakis and his colleagues, as well as by other researchers has produced more nuanced results – indicating when we should choose one method over another.[13] Similarly, the superforecasting project led by Philip Tetlock of the University of Pennsylvania has distilled good practice in political forecasting.[14] Big studies like these have helped Wharton School's professor Scott Armstrong to formulate what a good forecasting process should look like. As we saw earlier, Armstrong is a controversial figure. He's argued that the global warming forecasts of the United Nations Intergovernmental Panel on Climate Change (IPCC) don't pass muster, and says predictions that polar bear populations are heading for a dangerous decline should be dismissed.[15] Whether you agree with Armstrong or not, his conclusions are based on the extent to which he judges the forecasts of temperatures or polar bear populations measure up to a set of principles that are associated with accurate forecasting. He derived these by working with thirty-nine international experts in forecasting and had their findings reviewed by 123 other researchers. The principles have been regularly updated as new research evidence has emerged.[16]

The edited book that followed Armstrong's work has 139 principles. Some of these are based on common sense (e.g.

'test assumptions for validity'), while others have strong evidential support (e.g. 'describe reasons why the forecasts might be wrong'). A good forecast will survive these challenges and a poor one is likely to be improved as a result of them. Throughout, there's an emphasis on having a valid rationale for forecasts. As we have seen, getting forecasts to fit past data patterns is relatively easy and there's a temptation to think that a good fit is a sign of accurate forecasts to come. But, if there's no rationale for these forecasts, their performance is likely to be disappointing. We need to ask questions. Why do the forecasts predict a slump in sales every seventh week? Why do they predict an exponential rise in the number of cars on the road over the next few years?

We need to start placing a greater emphasis on judging forecasts by the quality of the process that was used to produce them. On its own, the past accuracy of a method in a particular situation will often be an imperfect basis for trust. And basing a decision on what a forecaster tells you just because they've previously made a few fluky forecasts that appeared to be accurate is almost a sure-fire way to be caught out.

# CHAPTER 11

# KNOWING WE DON'T KNOW AND THE DANGER OF KNOWING

## When we don't know

If we could travel back in time – even a few months – and tell people what was about to happen they probably wouldn't believe some of our prophecies. Around 2006, as Western economies motored along on their seemingly endless growth trajectories, I remember for amusement estimating how much the value of my house was increasing each day. I would have been shocked if I had known I would have to sell the house at a loss of £10,000 in 2010, following the financial crash. Forty years ago it would have seemed inconceivable to think industries that had dominated the British economy for centuries, like coal mining, would have all but disappeared by the dawn of the twenty-first century. And those who had worried about global cooling in the 1970s would be shocked to hear that global warming has become our main concern.

This world seems to be an unpredictable place, so can forecasting really play a useful role in it? Of course, not everything is unpredictable. The time the sun will rise tomorrow in London is perfectly predictable, although, for that reason, it's not interesting from a forecaster's perspective. But things like tomorrow's demand for electricity in the UK, consumer behaviour, the winner of the next US presidential election, crime rates in a district of New York and box office sales of the next Hollywood blockbuster have an enticing combination of systematic pattern and uncertainty that will raise a stir in a forecaster's heart. We don't know exactly how these events will turn out, but we have plenty of past data so we can attach probabilities to the different outcomes with reasonable confidence. In contrast, next week's prices on the stock market, the price of oil in a year's time and the state of the weather in Paris two years from now are largely unpredictable – but as we'll see, even here forecasting can have a role.

Of course, admitting that we don't know what will happen can take some nerve, especially if we've received huge amounts of funding to produce forecasts. The scientist James Lovelock wrote: 'I am proud to be part of a nation whose climate scientists had the courage in 2013 to say: We don't know'.[1] In fact, realising and admitting we don't know can actually be a valuable outcome of forecasting. It is much more useful than producing forecasts that give a pretence of knowing, such as 'we predict that a barrel of oil will cost $85 in a year's time' or 'the economy should grow next year by 2.8 per cent'. When we are aware we don't know it can motivate us to design a set of strategies that will allow us to survive or even prosper whatever the future throws at us.

Forecasting methods and tools when used correctly can still be helpful here. They can show us when there is no pattern in past events – that they occur entirely at random – even though we may believe we have spotted a pattern. They can reveal that many factors we thought had provided indications of what was about to happen in the future in reality had no predictive power at all. As we saw in Chapter 4, psychiatrists' assessments of the risk of violence posed by mental patients in New York had no association with their chances of being violent in the future. Peaks in solar activity, when there are more sunspots and geomagnetic activity, do not presage earthquakes.[2] The performance of candidates attending unstructured interviews is a poor predictor of how well they'll perform if they get the job.[3] Armed with findings like these we can be realistic and confront the fact that the future is more unpredictable than we thought.

Of course, we should always be aware that lurking in the forest there may be 'unknown unknowns', in the parlance of Donald Rumsfeld, the former US Secretary of Defense. This means we may not even know the bounds of possibilities or be able to list all the possible events – even if we do our best to imagine all the possibilities. Future technological developments fall into this category. We just can't anticipate all the technical marvels that might be a routine part of people's lives in fifty years' time. Natural events can also catch us out. In an exercise in October 2004, I asked over a hundred CEOs in Sri Lanka to list the factors they thought would impact on the tourism industry over the next five years. Their combined list was long and included factors like the possibility of civil war with the Tamil Tigers ending, cheaper flights and the likelihood of more visitors arriving from increasingly wealthy Asian nations such as China and India. Less than three months later the coast of the island was devastated by a factor that was not on anyone's list – the Boxing Day tsunami. Any forecast is predicated on the assumption that unknown unknowns do not exist. As long as we are aware of this we can trust the forecast for what it is – an honest, best assessment of the ways in which the future might unfold, based on our current knowledge.

## What to do when we know we don't know

There are several ways we can set about attempting to make ourselves future-proof when concerned about our lack of knowledge of what lies over the horizon. One established method is scenario planning. This comes in many forms, but it's claimed that the methods pioneered by Shell Oil helped make it the world's most profitable oil company in the 1980s. Although these methods originated in the business world, they can be adapted to help individuals create strategies that are intended to help them achieve their vision of the future in their personal lives.[4]

Scenario planning, as practised by Shell, is mainly about thinking in terms of extremes.[5] We identify the factors we think will have the biggest impact on the future but also have the greatest level of uncertainty associated with their future outcomes. We then write a series of stories that describe how the future might look if these factors turn out to be extremely good or bad.

For example, if we are a company considering manufacturing electric cars, the factors that would affect profitability over the next five years might include the percentage of purchasers who will opt for an electric car (market penetration), the costs of manufacturing the cars, the performance of the world economy and competition from other manufacturers. Of these factors we judge that two – market penetration and the future of the world economy – have the greatest level of uncertainty, and how they turn out will have the biggest impact on our future profits. We are very uncertain about market penetration as it depends on many things, such as petrol costs, future battery technology and how widespread the provision of charging points will be. The difference between extremely low and extremely high market penetration could mean the difference between bankruptcy and megabucks. Similarly, the state of the world economy over the next five years is highly uncertain. Whether it tumbles into a deep recession or achieves a high level of growth will determine the global demand for new cars and hence have a big impact on our profitability.

An example of a scenario which assumes that everything

turns out rosy is given below. Market penetration is high and the global economy grows significantly. Scenarios usually have names, so we've called this one 'Green Boom'.

## Green Boom

Greater world demand for oil as a result of strong economic growth pushes petrol prices significantly higher, making the running costs of electric vehicles relatively attractive. Environmental concerns over pollution trigger increased taxes on oil-based fuels in many countries. In most nations electric cars carry low or zero rates of taxation when bought and licensed. Developments in battery technology mean longer intervals between charging, while higher sales lead to significant falls in the manufacturing costs of electric vehicles. In addition, governments subsidise the development of infrastructures to support the charging of electric cars, including the widespread availability of power points in car parks and at supermarkets. This leads to high market penetration. World economic growth stimulates international demand so the size of the market for all vehicles grows significantly.[6]

We would write three other scenarios describing a future world where both factors turn out to be extremely negative – low market penetration and a world recession – or where one factor is extremely positive and the other extremely negative, assuming this is plausible (low market penetration and a world boom or high market penetration despite a world recession). It's very unlikely any of our scenarios will come to pass exactly as written. After all, they represent extreme outcomes. But between them the four scenarios represent bounds on where we think the world will go. The future world, we assume, will turn out to be somewhere between these extremes, but we make no pretence of knowing what this world will look like or which outcomes are most likely.

So what's the point of using scenario planning if it doesn't tell us what is going to happen? One answer is that it provides a 'wind tunnel' for testing how different strategies would fare in different futures. Ideally, we want robust strategies that would perform well in all the scenarios. If we can survive and prosper in all these extremes then we can be reasonably assured of our ability to cope with any future. Of course, designing such strategies is a non-trivial task. Some might perform well in one or two scenarios but disastrously in others. Good and robust strategies often have an element of diversity so that all our eggs are not in one basket. For example, a company that diversifies into tarmacadam manufacturing and consumer products would prosper if a government invested in road building during a recession to provide employment. On the other hand, if there's an economic boom, the company's consumer products would be expected to sell well.

Other benefits can result from using scenario planning. As humans we tend to assume the future will be like the present. The process of producing scenarios can jolt us out of this comfortable inertia and open our minds to the possibility of other futures.[7] It can prepare us for surprises and sensitise us to weak signals that indicate things might be changing. We are therefore less likely to ignore those dots on the horizon.

When groups of people meet to decide on future strategies, scenario planning allows them to understand and challenge each other's assumptions. In particular, minority opinions can be taken into account – they might just be correct. Winston Churchill was isolated politically when he warned about the dangers of Nazi Germany during the 1930s. In 2006, when most people thought that years of boom and bust in Western economies belonged to history, Peter Schiff, the president of investment firm Euro Pacific Capital, repeatedly predicted the 2008 financial crash on television. In Chapter 7, we saw that in forecasting we should distrust the person who deliberately sets themselves apart from others to attract attention, or the incompetent forecaster who just happened to be lucky last time when they made an extreme prediction. However, in scenario planning such views can be challenged. We can ask people to defend the underpinning rationale. If minority views survive these challenges they are probably worth taking seriously – they can represent another plausible future that we need to be aware of.

Scenario planning has seen many reported success stories. It prepared managers at Shell for the 1973 crisis, which saw a huge hike in the price of oil following the Yom Kippur War. Before the war there had been minimal changes in oil prices for over a quarter of a century and the vast majority of Shell's senior managers thought the price would remain at below $2 a barrel. But the company's scenarios included a world where oil was selling at the seemingly ridiculous price of $10 a barrel. This prompted Shell to make plans that would allow refineries to be converted rapidly, so that oil from other countries besides those in the Middle East could be refined. Within months the price had surged to $13 a barrel. But Shell's planning served it well. It was ready for the crisis and it changed from being the least profitable of major oil companies to the most profitable.[8] Later, in 1981, Shell sold off its oil reserves before prices collapsed following the outbreak of the Iraq–Iran war. The war had undermined the cohesion of the OPEC cartel that had managed to keep prices high. Lacking the benefits of Shell's scenario planning, rival oil companies made the mistake of stockpiling their reserves.[9]

Other companies who've had successes with scenario planning include Electrolux, which developed a set of scenarios relating to environmental issues. When asked to consider what they would do if they were living through a given scenario in ten years' time, Electrolux's managers began to realise that the components of the company's products could be recycled, and that doing so would produce financial and environmental gains. This led them to implement a successful strategy in their commercial cleaning business; they would rent their products to customers instead of selling them. The managers now saw themselves as supplying a service, rather than a product.[10]

We could go on. Through scenario planning Erste Allgemeine Verunsicherung, the Austrian insurance company, was ready for the fall of the Berlin Wall and this enabled it to move quickly into new markets in central Europe. KRONE, the wiring and cable supplier, developed 200 new product ideas after adopting the approach.[11] The peaceful transition to majority rule in South Africa was helped by scenarios that enabled the outgoing de Klerk government and the African National Congress to realise the importance of stimulating economic development during this period. In the light of these successes, scenario planning is becoming widely used. One survey[12] indicated that 65 per cent of companies it surveyed expected to use scenario planning in 2011.

## Stressed managers and causality

Of course scenario planning is not a panacea. Sometimes the scenarios make for uncomfortable reading and managers prefer to deny a potential scenario rather than confront a possible nasty outcome. Two British management researchers, Gerard Hodgkinson and George Wright, describe a case where they tried to apply scenario planning to a company and failed.[13] The company was threatened by new technology which could make its current business offering obsolete, but the attempt to use scenario planning served only to bring difficult issues to the surface. The company's domineering CEO attended the early meetings squeezing a soft sponge, apparently to relieve her stress, and at some meetings she became agitated and paced up and down. Later she announced that she would not be attending subsequent discussions, arguing that the process would not deliver any useful ideas or insights – despite other members of the management team seeing value in the exercise. The researchers concluded that stresses arising out of attempts to apply scenario planning led the team to resort to a series of coping methods. These included finding a rationale to bolster support for the company's current strategy – even though this was evidently threatening its very existence.

Then there's the issue of causality.[14] Scenarios emphasise causal chains that lead us from where we are now to a particular, extreme world. 'World economic growth leads to more demand for oil which leads to increases in its price so electric vehicles become relatively cheap to run and sales skyrocket'. This is consistent with the belief that every future event has a cause. In this view, attributed to a paper published in 1814 by French scholar Pierre-Simon Laplace, our uncertainty about what will happen in the future arises because we have a problem identifying and measuring causes. According to this position, if we knew the current state of everything in the universe and could identify and measure every force that determined their future states then all things – from the movement of atoms to world events – could be predicted with absolute certainty. Armed with this knowledge we could even predict the outcome of the toss of a coin or the throw of a die with perfect accuracy. This has been referred to as scientific determinism.

Laplace died in 1827, long before the curious and mind-bending discoveries of quantum mechanics such as quantum entanglement, or the finding that subatomic particles can be in more than one place at a time. Discoveries like these led to an alternative view of the origins of uncertainty. They suggested that events can arise purely through chance. The same set of factors can be in place on different occasions yet they can be associated with completely different outcomes. In these cases our search for causality is misplaced. No amount of knowledge will allow us to say with certainty what the next event will look like. Some people have argued that, even when events do have causes, if we are unable to identify what these causes are then we may as well regard the events as random.[15]

Scenario planning acknowledges uncertainty by generating a range of scenarios without attempting to make an assessment of what will actually happen. But each scenario makes no mention of uncertainty. They individually describe neat worlds where events assuredly follow on from their causes. This can make the small number of scenarios appear highly plausible and lead people to develop an unfounded belief that the chain of events to which they relate will actually occur. Consider these two short scenarios.[16]

*Scenario A*
The US computer manufacturing industry experiences a decline in its share of the global market during the third decade of the twenty-first century.

*Scenario B*
The US computer manufacturing industry experiences a decline in its share of the global market during the third decade of

the twenty-first century as a result of competition from Asian countries such as, China, Japan, Malaysia and South Korea.

Which scenario is most likely to prevail? The problem is similar to that of predicting whether John will make a loss on his market stall sales, as we saw in Chapter 1. Many people would opt for Scenario B. It seems more plausible because it gives a causal explanation for the declining market share and it's easy to imagine this scenario unfolding. But Scenario A must be more likely. It describes a situation where the declining market share can occur for any reason – not just the specific reason given in Scenario B. Even expert forecasters can make this mistake. In 1982, two famous psychologists, Amos Tversky and Daniel Kahneman, asked a group of delegates attending the Second International Symposium on Forecasting in Istanbul to assess the probability of the following event occurring:

A complete suspension of diplomatic relations between the USA and the Soviet Union, sometime in 1983.

A second group was asked to assess the probability of the following event:

A Russian invasion of Poland, and a complete suspension of diplomatic relations between the USA and the Soviet Union, sometime in 1983.

Both groups typically estimated probabilities that were low. But on average they judged that the second specific event was over three times more probable than the first one.[17]

Ironically, as more and more detail is added to a scenario, the more plausible it can appear, yet its actual chances of occurring decline. People can become unrealistically attached to the small set of scenarios they've produced so they are immune to the prospect of other scenarios occurring. Unless they are careful this will narrow their perspectives and make them vulnerable to shocks and surprises.

There's another reason why scenario planning's emphasis on causality can make us comfortably blinkered to the possibility of shocks. I used to teach the topic to the employees of a large science-based company that had major defence-related projects. For this company a positive scenario was a world where there was a rise in conflict, crime and terrorism. In a world of peace and harmony no one would be interested in the company's highly sophisticated products that scanned people efficiently at airports or provided enhanced cyber security. But one question that frequently came up was: where do we draw the boundaries of possibilities in our scenarios? For example, should they contain the possibility of an invasion by extra-terrestrials? We might be crazy to consider this, but isn't scenario planning about preparing for all eventualities? Isn't it about 'thinking outside the box' and not being shocked by events no one else thought were possible? The answer is that the scenario should only contain a possible invasion from outer space if we can find a causal chain linking our present circumstances to that event. It's highly unlikely we could find such a causal linkage. We are not in dispute with extra-terrestrials over mineral rights on Mars or violations of their privacy on Saturn's moon Enceladus, so we are unlikely to receive an ultimatum that threatens invasion. This may sound silly, but it illustrates a serious issue. Scenario planning cannot prepare us for surprise events that arrive out of the blue with no apparent cause.

So can we confront the future without having to think about causality? In some situations we may have no choice. We have to accept that we have no idea what is going to happen and what, if anything, will cause the events that will emerge. We may be unable to produce an exhaustive list of possible events, but we think that something major and unforeseen – good or bad – may happen. Like a person in the jungle who is frozen with fear because they sense an unidentified predator might be watching them, hidden by the tangled vegetation, our attempts to envisage the future can be paralysed by the lurking 'unknown unknowns'. What can we do?

### Avoiding fragility

Nassim Taleb, the acerbic former trader and now a hugely successful author – he was described in *The Times* as the 'hottest thinker in the world'[18] – suggests we should attempt to make ourselves 'antifragile'.[19] Fragile items, such as china plates or mobile phone screens, shatter when they are subject to stress or adverse conditions like a sudden strong force. In contrast, antifragile phenomena become stronger through adversity. Our immune system is strengthened by exposure to bacteria and viruses. Our thinking and ideas often improve when exposed to challenge and debate. Children become emotionally stronger if they are not overprotected from the knocks and scrapes of everyday life. If they're allowed to, managers can learn from their mistakes and improve their skills – though unfortunately some organisations operate a 'one big mistake and you're out' policy.

While fragile items can withstand many small knocks, one large bash will destroy them. A heavy hammer blow to a plate will do far more damage than a hundred small taps that cumulatively add up to the same force as the hammer blow. For an individual or an organisation the danger is that a sudden surprise event will occur with the destructive power of the hammer blow and overwhelm them. We may be able to predict small taps with reasonable confidence, but the hammer blow may appear to come from nowhere. To survive and thrive in this world of dangerous and unpredictable large shocks – so-called black swans – we must try to organise ourselves so that any potential losses we might experience in the future are limited, but potential gains are unbounded. Examples include investing 90 per cent of our savings in relatively safe assets, such as government bonds and risking the remaining 10 per cent in highly volatile investments. We can't lose more than 10 per cent of our funds. But we just might make huge profits.[20]

Notice that antifragility is all about putting ourselves in a position where we might gain substantially from the shocks and surprises of the future and, at the very worst, we will not be wiped out. It doesn't require us to identify and understand the causes of events. It assumes that what is to come is unknowable. As such, it adopts a completely different perspective than that of the super-predictor envisaged by Laplace.

Can Taleb's antifragility ideas be put into practice? James Derbyshire and George Wright of Anglia Ruskin University and the University of Strathclyde have attempted to translate them into a series of formal steps.[21] These involve asking whether a proposed course of action will make us or our organisation vulnerable to great harm in the future or whether there is an acceptable cut-off point to potential losses. If there appears to be fragility in the proposal, Derbyshire and Wright recommend considering whether it can be made more antifragile. This might be achieved in a number of ways. Building flexibility into plans allows us to adapt to unforeseen changes. For example, negotiating a 'get out' clause in a contract would allow us to avoid further commitment if things should go awry.

Building redundancy into a system can also provide more resistance to adverse future events. For example, personal savings will help to mitigate the consequences if the roof of our house is blown away in a tornado. In fact the tight efficiency of some systems is inimical to antifragility – one small hitch in part of the system can result in massive problems. A take-off-and-landing schedule at an airport that seeks to use the runway at maximum capacity could result in chaotic delays if just one aircraft arrives a few minutes late.

Finally, tinkering and seeking to improve things in small

steps by trial and error will allow any mistakes to be rectified
at relatively little cost. That way we won't incur the risks associated with giant leaps. As Yale Professor Charles E. Lindblom
argued in his 1959 article 'The Science of Muddling Through',[22]
governments can avoid the dangers of large-scale policy
changes by making small tweaks that are intended to enhance
current policies – raising the pension age by a year or closing
a tax loophole, for example. Fundamental overhauls of the
pension or tax systems could turn out to be disastrous – think
of Margaret Thatcher's poll tax. Similarly, organisations might
offer a new service in just one small locality before gradually
rolling out the service to other parts of a country, if things go
well. Or individuals might test out a possible new career – like
becoming a novelist – by doing increasing amounts of similar
work in their spare time, before gradually moving to a career
change if their initial experiences prove successful.

## When we don't want to know

While we might be completely uncertain about the future in
some cases, in others we can make forecasts with a high degree
of certainty – but is this always a good thing? Huntington's
disease is a genetic disorder that leads to severe mental decline
and a deterioration in muscle coordination. Symptoms, such
as jerky movements, unsteady walking and subtle psychiatric changes, typically first appear when people reach their
mid-thirties or early forties, but they become progressively
worse. The disease is particularly cruel because the children

of those affected have a 50 per cent chance of inheriting the condition.

American singer-songwriter Woody Guthrie died of the disease aged only fifty-five and, in 1968, his wife, Marjorie, founded the Huntington's Disease Society of America. Among its many activities, such as providing support for families affected by the condition, the society funds research that aims to find a cure. Although this has yet to be achieved, advances in genetic science in the 1990s led to an inexpensive test that predicts with certainty whether or not a person will develop the disease. Yet in the USA only 5 to 10 per cent of at-risk people opt to take the test.

This puzzled Emily Oster, an economist at the University of Chicago, and her co-researchers. Economists' models usually assume that accurate information about the future – as long as it can be obtained inexpensively – is always a good thing since it leads to better decisions. They decided to investigate why so few people were prepared to take a test that would remove all the uncertainty about their future prospects.[23] In a study that involved interviewing over a thousand individuals who were in danger of developing the disease they found that people simply did not want to live with the anticipation their future health would severely decline while they were still relatively young. If they did not take the test they could continue to believe they would stay relatively healthy, regardless of their true state. Overall, the study found that untested people tended to be overly optimistic, especially when symptoms consistent with the disease became more evident. When these symptoms indicated a 100 per cent chance that the disease was present their personal estimate of the probability was just over 50 per cent. Taking the test could destroy this optimism. Ironically, the researchers suggested that making the test less accurate might make it more attractive. A positive result would still allow for the possibility that a person was disease-free but it could also prompt them to make contingency plans for the future in case the result proved to be correct.

### Police, movies, music and a dodgy car radio

Predictive policing is another area where reliable forecasts can create problems, despite their obvious benefits. Police forces in areas such as Los Angeles in the US and the county of Kent in the UK now employ sophisticated algorithms to predict crime hotspots based on big data. Officers in patrol cars can watch screens where red dots indicate locations, such as car parks or blocks of flats where the probability of crime is high. Some algorithms even take into account the weather – muggers don't like getting wet – and whether today is a payday when they make their predictions. As a result, police resources can be used more efficiently and crime rates reduced. In the Foothill district of Los Angeles the introduction of predictive models in 2011 reduced property crime by 12 per cent in its first six months, compared to the previous year. Adjacent districts, which had conventional policing, saw a rise of 0.5 per cent in the same period.[24]

But the way these accurate forecasts are used can also bring potential problems. An area predicted to be a crime hotspot might lead to overzealous stopping and searching of people, leading to a vicious circle of alienation and more crime. Andrew Ferguson of the UDC David A. Clarke School of Law in Washington DC has additional concerns.[25] The Fourth Amendment of the US constitution prohibits unreasonable searches, seizures and arrests and asserts that these must be supported by 'probable cause' or reasonable suspicion. But, argues Ferguson, suppose that the predictive model indicates there is only a 5 per cent probability a car crime will be committed on a particular street at a given time. Yet, when patrolling the street, police officers spot a young man holding a screwdriver and standing beside a parked car. Their first reaction is that a car break-in has been committed, so their natural instinct is to stop and search the potential offender. But given the low probability indicated by the algorithm, do they have probable cause to do this? Could the Fourth Amendment be used against them?

Ferguson also worries that courts may place too much emphasis on the probability forecasts of predictive policing algorithms because they are objective. If a model predicts that 90 per cent of men on a street at a particular time are carrying drugs and a man arrested on that street appears before the court, it may be difficult not to be over-influenced by this high probability. The danger is that it might eclipse other significant evidence when the court makes its decision.

It's not only in the unpleasant things in life, like disease and crime, that reliable forecasts can be a mixed blessing. In 2004 Ben Novak was a struggling musician hoping to break into the big time. As he drove around his native New Zealand, Ben's main complaint about his eleven-year-old Nissan Bluebird was the radio. He could only listen to two stations – an annoyance to someone who lived and breathed music. Yet this restriction was to be the making of Ben.[26] One of the stations he had access to was the BBC and one day they carried a short item about an algorithm which claimed to be able to predict the songs that would be hits. When he got home Novak found the algorithm's website and, on the off chance, uploaded a song he'd written a few years earlier called 'Turn Your Car Around'. The analysis cost him $50. The website produced a numeric estimate of the song's hit potential – anything scoring more than seven had a very high chance of rocketing up the charts. Novak's song was rated at 7.57, a score similar to some of the charts' all-time biggest hits, like Steppenwolf's 'Born to be Wild'.

Over in Spain, staff at the company that produced the algorithm (then called Polyphonic HMI) were alerted to the fact an uploaded song had achieved such a high score. The company's boss, Mike McCready, circulated the song around record companies in Europe. Within a couple of weeks a successful music producer was offering Novak a deal: upcoming pop star Lee Ryan needed a new song and 'Turn Your Car Around' would be just perfect. If Ben agreed, he'd get 50 per cent of all royalties. The result: a song that for two consecutive months was the most played in Britain.

A great outcome for all concerned – but are predictive algorithms like this really good for the creative arts? Nowadays computers can analyse and assess prospective film scripts and reliably predict box-office takings. Ever since a company called Epagogix analysed the scripts of nine soon-to-be-released films and made spot-on predictions for six of them, algorithms have become a key decision-making tool in Hollywood. Analysts at Epagogix, operating in the attic of a business park in London, score a film's scripts – everything from gunfights to love scenes – against a list of criteria including proposed actors and locations. They then feed their ratings into the computer, allowing the algorithm to use its huge database of scores and takings for previously released films to produce its prediction.

Given its apparent accuracy, some movie financiers will only back a movie if it receives a positive prediction from Epagogix. The danger is clear. Those risky movies that offer a complete break from convention might never get off the ground. There's a temptation to play it safe and stick to a tried and tested recipe for success. To some extent the forecasts will be self-fulfilling in their apparent accuracy. We will never know whether the rejected films would have been a success. For both music and movies the excitement of new idiosyncratic and revolutionary creations that drive progress in the arts might be supplanted by dull formulaic standardisation based on what worked in the past. The danger is much more serious for films than music as they usually require huge financial backing. These days new music can be produced on a shoestring and broadcast to the world from a tiny bedroom, though it might not get the backing of big promoters.

## Would you buy a perfect prediction machine?

Suppose I invented a machine that could accurately anticipate the result of every decision you make. It won't predict the rest of your life – I'm not that clever – but it will tell you what will happen one decision at a time if you make a particular choice. It won't cost very much. Would you want to buy it?

It might save you a lot of money. When you ask: 'What will happen if I buy a lottery ticket today?' It's virtually certain to tell you that you won't win. Sometimes it might surprise you. No, you won't enjoy that holiday in the Caribbean. Yes, you will get that great job hundreds of people are applying for. It seems a no-brainer – buy the machine. But there's one condition that comes with the purchase – you must use it for every significant decision in your life. That way you'll never get a decision wrong. Or will you?

Suppose my machine had been available back in 1956 to Marc Chavannes, a Swiss inventor, and Alfred Fielding, an American engineer, who were trying to create a new type of trendy wallpaper working in a garage in Hawthorne, New Jersey. Their idea was to create air bubbles between two sheets of plastic to create an attractive pattern. If they'd asked my machine whether the new type of wallpaper would be a success, it would have told them they were heading for abject failure. They would probably have abandoned their dream. But negative outcomes can sometimes lead to huge successes. What proved to be a useless idea for wallpaper turned out to be a brilliant idea for packing material – bubble wrap. Around $400 million worth of the material is now sold annually.

By not knowing what will happen in the future we can make the wrong decision. But as in the case of bubble wrap, mistakes can open up new opportunities or help us to avoid even bigger disasters. In China it is widely believed that good things often come from bad events. Referring to an ancient story, there is a Chinese saying: 'Who could have guessed it was a blessing in disguise when the old man on the frontier lost his mare?' In the story the old man's loss of his horse seems to lead only to more misfortune when his son breaks his leg. Yet the son's injury saves him from having to join the army in a war where most young men are killed.[27] If Ben Novak's car radio hadn't been defective, he might never have experienced the big time. And, as we saw, Taleb's antifragility concept suggests that if we stumble into adversity we can often gain from it and grow.

Of course, this is not a manifesto for setting out to make mistakes. It's sensible to always aim for what appears to be the best decision. And forecasting can often help us to go a long way towards getting our decisions right. But even trustworthy forecasts can't guarantee that our choices will lead to what we immediately want. The known randomness and the unknown unknowns are likely to be beavering away to create their shocks and surprises. But what would life be like without them? American poet Alice Walker wrote: 'Expect nothing. Live frugally on surprise.' And Shakespeare told us that 'All the world's a stage'. Sometimes our lives can seem like dramas in which we have a starring role. We may do our best to ensure things end happily, just as we intended, and often we'll be successful. But without shocks and surprises would we be happy? And as our faultless performances progressed smoothly from act to act wouldn't they be witnessed by tier after tier of silent and empty seats?

# CHAPTER 12

# CONCLUSIONS

### A road to prediction

One day, when I was seventeen, I knocked timidly on the door of the teachers' staff room at my school. It was lunchtime. Even in the corridor outside, the air was thick with the smell of cigarettes and a hubbub of laughter and muffled male voices made me feel I had no right to be there. It was as if I was interrupting a private party. Mr Haden, the teacher I wanted to see, opened the door still wearing his gown. He glared at me through rimless spectacles; his protruding front teeth seemed to emphasise his obvious distaste at seeing me there. He didn't speak. I felt even more uncomfortable.

'Please sir,' I said, 'could I have one of those university application forms that you talked about this morning?'

Mr Haden's mouth tightened around his teeth. He let out a sigh as if I was asking him for a large loan. I sensed he was turning over phrases in his head to work out which one would knock me back the furthest.

'Is there any point in you applying to university, wasting your teachers' time filling in references for you? University places are reserved for the elite. You don't have a snowball's chance in hell of getting in.'

I knew Mr Haden had a first-class honours degree from a top university and was known to be very clever indeed. And he delivered his verdict on me with such authority that, for a moment, I wondered whether people like him had mysterious predictive powers. Perhaps he could infer from the shape of my head or the timbre of my voice that I was without doubt a no-hoper. In truth, he hadn't taught me for several years and he probably still regarded me as the frivolous and easily bored fourteen-year-old who'd paid little attention in his classes – more interested in football and rock music than French irregular verbs and quadratic equations. But I'd worked hard in the intervening years.

Well, somehow I managed to get hold of a form and, partly galvanised by Mr Haden's pessimistic prediction, I worked even harder and did get a place. Nearly forty years later the memory of this incident came back to me on the eve of the inaugural lecture that I was to give as a new professor at my university. The lecture was titled: 'Improving the Future of Forecasting' and I found myself reflecting on the power of predictions to do harm or good. Mr Haden's casual pronouncement, which was based on out-of-date information, yet delivered with supreme confidence, could have damaged me for life. Fortunately, it achieved the opposite effect. It was enough to make me sceptical about predictions.

My university place was at Liverpool, then one of the trendiest cities on Earth, to study the trendiest subject of the late '60s, economics. It was an age of optimism – and economics was surely the key to making the world a better place. After all, Keynesian policies had eradicated the horrors of mass unemployment – the despairing stagnation of the 1930s belonged only to a cold, far-off age. We had witnessed year after year of economic growth – shiny new consumer goods, assertive brutalist architecture, a Prime Minister promoting the virtues of white-hot technology, men on their way to the Moon. There were no limits. Of course, there was still poverty and conflicts still existed both at home and overseas, but these problems invariably had economic roots and, as our understanding of the subject advanced, they too would be confined to history books. The title of a contemporary economics textbook summed it up: *The Science of Wealth*.

And what a place to be. 'At the present moment,' declared the American poet Allen Ginsberg, 'Liverpool is the centre of consciousness of the human universe.'[1] Here was the home of The Beatles and the world-renowned Mersey Sound. It was where the country's top entertainers had honed their skills, infused by the city's raw seaport culture. This was a big, confident and unpretentious left-wing city where on a Saturday afternoon deafening crowds squeezed on to the Kop to cheer for the best football team in Europe. And it seemed natural to be studying economics in the north. To me economists with northern accents had a certain cachet. It was as if their ideas had been forged in the hard, no-nonsense world of the mine and mill – their streetwise, 'call a spade a spade' attitude giving short shrift to romantic ideals and self-indulgent intellectualism. Robust practical ideas that might help to solve the world's remaining problems – that's what I expected to be discovering as I unpacked my cases on my first day on Merseyside.

Alas, for me and economics, and for Liverpool in the medium term, unrealistic optimism had prevailed. Liverpool is westward facing. It was designed for the transatlantic trade of the cotton manufacturing era. Modern Britain was looking eastwards, embracing business with Europe through hi-tech container ports. Before long, the city was to endure the agony of large-scale unemployment, poverty and deprivation. In 1981, the Toxteth district, where I'd lived, was torn apart by riots, rooted in these deep social problems. Some people even questioned whether the city was an anachronism.

And economics? Far from the pragmatism I'd expected, my time was spent contemplating strange abstract worlds. In these worlds the denizens never departed from perfect rationality. Their brains had the capacity of computers enabling them effortlessly to process all relevant information and to perform heroic calculations so that, unfailingly, all their decisions were optimal. Each world was built on layers of convenient assumptions, like a house of cards, and clever mathematics was then applied to predict the inhabitants' consistent, self-interested and emotionless actions. There's a well-known joke about an economist, who's marooned on a desert island with a physicist and a chemist, and only a tin of baked beans to sustain them. Frustratingly, they find they have no tin opener. The economist is puzzled by the discussion between the other two about how the tin might be opened. 'But why don't we just assume we have a tin opener,' he says, in the happy belief that this will solve the problem.

I was studying a branch of mathematics, not a social science. The imaginary mini-worlds, propped up by unrealistic assumptions, were elegant intellectual puzzles – doubtless excellent devices for training the analytic brain – but they did not satisfy my yearning for practicality. In hindsight, all those bold, youthful hopes and predictions seemed like a worthless dream.

There was, however, one aspect of the course that had practical appeal – statistics. Statistics can yield information and knowledge. And, as Sir Francis Bacon allegedly said, '*scientia potestas est*' or 'knowledge is power'. Statistics can save lives – they can uncover links between disparate lifestyle factors and diseases. They can expose fraudsters. They can reveal to agriculturalists the secrets of how crop production can be enhanced. They can help sport coaches to engineer the defeat of opposing teams. And they can help companies to make more informed decisions.

So, sometime after graduation I found myself near the east

coast, in Hull, teaching statistics at evening classes to managers in an old Victorian school that had cracked blackboards and a nearby tidal river that occasionally threatened to flood the premises. This was the era of management institutes. There was the Institute of Works Managers, the Institute of Administrative Managers, the Institute of Personnel Managers, the Institute of Work Study Practitioners and the Institute of Management Accountants. They all held examinations in statistics and the managers, despite often being tired after a day's work, were keen to learn. Sometimes they would bring me statistical problems from their place of work. And one problem came up often: 'how do I forecast my sales, my costs, my staff turnover or my raw material requirements?'

My interest in prediction was ignited. This was practical stuff. It wouldn't change the world, but it could make a small difference, helping people to manage their organisations a little better. It was fun to see how well the forecasting methods we worked on fared when the real figures came in, and rewarding when they appeared to be working efficiently. This was especially true when they outperformed the judgemental forecasts of the managers. The statistical methods didn't always come out on top. Sometimes the managers' forecasts revealed a long-held belief in patterns that weren't there, and their forecasts suffered. But, on other occasions, their insights into the mysterious ups and downs in their graphs proved spot on. Making forecasts was about pitting yourself against the caprices of human behaviour and nature, rather like a sailor and a boat taking on the challenge of the Roaring Forties. And the best chance you had of winning often came from a combination of management expertise and statistical analysis.

The role of human judgement and expertise in prediction intrigued me. Economics had taught me that people making predictions would use their past experience and all available information rationally and optimally. As John R. Lott Jr., the controversial American economist, once commented to me after I'd presented at a conference in New York, if they didn't, the ruthless power of the market would surely drive them out of business. But here there were examples of people misreading information, making unwarranted assumptions or ignoring highly relevant signals about the future. Yet their organisations persisted and even prospered. Perhaps it was the benevolent corporate environment of the '70s – pre-Margaret Thatcher. Or perhaps economics hadn't quite got it right.

When a colleague introduced me to the work of psychologists Amos Tversky and Daniel Kahneman, it was a revelation. Their papers were full of clever experiments that revealed, in terms readily accessible to those not trained in psychology, the cognitive biases inherent in our judgements when we're confronted with uncertainty. So that intellectual superhuman, *Homo economicus*, was a myth after all. Here was evidence that people are prone to seeing false patterns in random events or over-relying on their fallible memories when making judgements. Experiment after experiment revealed how we could be over-influenced by irrelevant numbers planted in our heads, and that we have a propensity to favour unreliable, colourful anecdotes over hard statistical evidence. Despite (or more likely because of) his myth-busting conclusions, Kahneman went on to win the Nobel Prize in Economics in 2002. Sadly, Tversky who had contributed so much, had died in 1996.

Soon I'd enrolled for a PhD, aiming to discover how the biases revealed by Tversky and Kahneman played out in forecasts made in organisations, and whether they could be countered. I was soon to find that it's easier to discover biases than to prevent them. For example, I found that managers often forecast too high when their sales were low and too low when their sales were high. I developed a mathematical method for correcting the problem. You simply fed their forecast into my formula and, hey presto, the biases were removed. The formula could even learn and adapt if people's biases got better or worse. But there was a problem – if managers heard that their forecasts were being corrected they might treat them less seriously. Why bother, if you knew your forecast would be changed anyway? Worse still, an arms race could develop with the manager deliberately escalating their forecast to try to pre-empt any correction that they thought might be applied.

It's challenges like these that make forecasting interesting – whether it's how we form our personal expectations about the future or how organisations make their forecasts. Even so, it's odd that, despite my early scepticism about prediction, I ended up working in forecasting – consulting for companies and government departments, researching forecasting methods, helping to edit a forecasting journal and teaching students the very latest prediction techniques. On the way, I've developed a tolerance for the imperfection of prediction and its inevitable suboptimality. There's even enjoyment to be had in the battles between the neat formality of forecasts and the untidy world they seek to tame. The world's a wild and messy place and we'll seldom find solutions to problems that are truly optimal. We have, instead, to compromise and approximate, to guestimate and arbitrate and hope that what we achieve is at least satisfactory and worth the effort involved. Above all, like that yachtsman battling with the giant waves of the Roaring Forties, I've learned that a good forecaster needs a humble respect for nature. Without that, the future can appear like a leviathan to cast you on a rocky shore and pound your cherished plans to smithereens.

## An end to prediction?

It's easy to understand why forecasters often display a lack of humility – as consumers of forecasts we tend to demand confidence and certainty from those making predictions. Otherwise, we close our ears. A few forecasters may be genuinely deluded that they can see the future laid out with the accuracy of the latest satellite photograph. But feigning certainty where you know none exists is being economical with the truth – it's lying and potentially harmful. Better to tell the truth and not be listened to than to tell an untruth and cause a catastrophe.

In his bestselling book *The Black Swan*, Nassim Taleb

claims that 'some forecasters cause more damage to society than criminals'.[2] He is undoubtedly right, but there are also some forecasters who do a tremendous amount of good. In the USA, weather forecasts have been estimated to be worth, on average, $286 per year to each household.[3] An independent review[4] of the value of public weather forecasts in the UK concluded in 2015 that they brought economic benefits of between £1 billion to £1.5 billion and that weather warnings may be saving 'many tens of lives each year from direct impacts of the weather', such as floods and heatwaves. A study of the economic value of weather forecasts to companies in South Korea and China suggested that higher profits could be obtained by using the forecasts – as long as decision-makers did not underestimate their general reliability because they recalled the occasional inaccurate prediction.[5] Flood forecasting and warning systems in Europe have been estimated to bring benefits worth €400 for every €1 invested.[6]

It's frustrating to find a bare shelf in the supermarket where your favourite cheese or bread should be. If this rarely happens you can probably thank the company's forecasters for accurately anticipating the demand for the product. Of course, the supermarket may have kept the shelves full simply by deliberately overestimating demand. The result can be bins full of unwanted food hidden away at the back of the store awaiting transfer to a landfill site – a tragic waste in a world where resources are becoming scarcer and millions don't get enough to eat. To counter this some supermarkets are using sophisticated systems that predict our changeable tastes and demands. For example, they know we tend to buy more salad on a Monday having stuffed ourselves at the weekend. But it's our reactions to small changes in the weather that these systems are particularly clever at anticipating. Tesco have found that burger sales will surge by 42 per cent when the thermometer rises from 20°C to 24°C. When a summer weekend sees a ten-degree rise, coleslaw sales will rocket by 50 per cent, but fewer people will want green vegetables. And, in case you want to know, when a hot spell starts at the weekend, sales of hair removal cream can increase by as much as 1,400 per cent as women switch from jeans and tights and expose their bare legs. Then, as soon as the weather turns cold, people's trolleys fill up with soup, long-life milk, sausages and root vegetables.[7]

Not all products are as easy to predict. I once attended a forecasting meeting at a leading food company where the managers were scratching their heads trying to forecast the sales of a new and 'funky' purple-coloured tomato sauce. The product was being launched in the hope that parents would soon be queuing up to buy the product to please their kids (green, pink, orange, teal, and blue-coloured varieties were also available). Thoughtfully, Vitamin C was added to the sauce to please the parents. The actual demand for coloured ketchup surged for a short time: 25 million bottles were sold in the US alone – but the novelty wore off and, by the time the product was withdrawn a few years later, kids were pestering their parents in the supermarket aisles for something else.

Nevertheless, accurate predictions can help to ensure that the goods we need are available when we want them – and without the cost of rubbish dumps stuffed with expensively produced and expensively delivered waste. Accurate prediction is good for the environment.

There are many other examples of where accurate predictions benefit us. The danger of the assertions made by people like Dominic Lawson – who you will recall from the Introduction cheerfully predicted an end to prediction – and Nassim Taleb is that they might give the impression that, because forecasts of exchange rates, commodity prices and stock market prices are often misleading (not least because they fail to acknowledge uncertainty), then all prediction is a dangerous alchemist's dream. As John Van Reenen, professor of Applied Economics at MIT, argued: if you believe that then people should ignore their doctor's advice to give up smoking because the medical profession failed to forecast the Aids epidemic. Imagine a country where an authoritarian government made prediction illegal; where people could be thrown in jail for the heinous crime of estimating what might happen in the future. All right, there would be no celebrity weather forecasters cracking the hackneyed joke that they're not to blame for the latest deluge. And there'd be an end to the hours of empty speculation by TV sports pundits that's designed to fill the schedules cheaply and keep the accountants happy. For better or worse, the gambling industry would disappear. But there would also be chaos in shops. We'd see long queues reminiscent of Cold War Eastern Europe as rumours spread that the latest must-have product had made a rare appearance on the shelves. There'd also be little or no development of new services or products. After all, which airline would be prepared to consider investing in a new route if it had no forecasts of the likely demand? And who'd be prepared to finance an investment in an innovative gadget when they were not allowed to predict what its sales might be? We would probably have much less time to use these products and services anyway. We'd spend even more hours on the phone awaiting a response from call centres than we do now because the centres' managers would have no guidance on how many operators they needed at any given future time. And if a new gadget needed electricity – beware. Electricity companies would struggle to match their generating capacity to future demand.

Much worse, many people would die. There'd be no guarantee that roads would be salted on winter nights when ice turned them into mirrors. Residents living near rivers could not be told when a flood was an hour away from engulfing their homes. Or they'd be evacuated regularly when there was no real threat of flooding. Hospitals and police forces would struggle to match their staffing levels to needs. And, ironically, the government that brought in the ban would find itself sailing its macroeconomic ship without any idea of where it might be heading.

The answer is not an end to prediction but an end to bad predictions, and we've seen plenty of the latter in the previous chapters. So how do we separate the good from the bad so that, when we use predictions to make decisions, we can be confident we are employing the best guidance available? Unfortunately, when a prediction is presented to us, and particularly when we have no idea of the track record of the predictor, there is no acid-test to tell us whether to trust it or not. But the bad prediction is not completely undetectable. Like an unsolved crime it comes with a pattern of evidence, clues and plausible motivations. To try to apprehend a bad 'un before it does us harm, let's distil the evidence of the earlier chapters. If we do, we can identify three key traits that might give it away: contamination, incompleteness and ineptitude.

## Contamination

Predictions are contaminated when they are not an unbiased assessment of what will happen in the future, but are designed, instead, to serve some other end. We must be wary of those more interested in their reputation than their accuracy. These include the herders – who, you may recall, simply agree with other forecasters despite their own beliefs because they seek the safety of the crowd in case their forecast is wrong. Look out also for the anti-herders. They hope you'll remember their occasional lucky forecast that was different to everyone else's, but forget the numerous times they were wrong. Then there are those with other agendas – politicians and lobbyists, publicity seekers and hack journalists, those with vested interests and market manipulators – and junior staff distorting their view of the future to please the boss. It's always a good idea to ask: does this prediction just happen to suit this person or group? If it does, we need to be vigilant. Demand evidence and a rationale and, if possible, test its plausibility compared to that of other possible predictions. Above all, ask whether the forecaster is likely to have any interest at all in producing reliable predictions.

## Incompleteness

All forecasts will be incomplete – no prediction can encompass all the possible ways that the future might unfold. Even when you predict the outcome of a coin toss, there's a possibility that the coin might be snatched in mid-air by a skilful thief, make a perfect landing on its rim, disappear through an open window or be subject to an unimaginable number of other bizarre events. More serious are the black swans, made famous in Taleb's eponymous book. These are unforeseen and rare events that can have massive, and in some cases catastrophic, impacts when they occur. Massive stock market crashes, off-the-scale floods and tsunamis, huge terrorist attacks, disruptive technological innovations, military coups, even new fashions – all potentially lie in wait to shock and surprise us, just as the discovery of avian black swans astounded the first European arrivals in Australia, convinced as they were that all swans were white. Events we have never witnessed before and that we have no way of contemplating are bound to be absent from our forecasts – the best we can do is try to cover ourselves against the unexpected, as we saw in the previous chapter. When we see a forecast it's wise to bear in mind this incompleteness. But surely it doesn't mean that we should abandon prediction. I drive a car each day implicitly predicting that I'll reach my destination safely, but there is always – a hopefully tiny – chance that disaster will ensue. But if I let such worst-case scenarios rule my life I would never drive; I wouldn't go on vacation, or try new foods ... in fact I probably wouldn't have a life at all. Imagine telling a food manufacturer to stop forecasting sales, because there's a remote chance that a future TV documentary will uncover evidence that the manufacturer's product is carcinogenic causing that nice, regular sales graph to plummet to zero like a meteor. It's wise that the company should have plans in place to deal with a disaster. But it should also carry on forecasting in the meantime.

The less justifiable form of incompleteness is the single number or single event prediction that, like a fig leaf, attempts to hide the uncertainty behind it. We saw in Chapter 8 that we tend to prefer these neat, precise and relatively unchallenging predictions. Like junk food, they are palatable and make us feel comfortable. But, just as junk food is bad for our waistlines, these abbreviated forecasts are likely to be no good for us when we have to make decisions. When the weather forecaster says it will be fine and sunny tomorrow I still need to know the risk that I'll arrive at my destination soaked to the skin if I decide to walk there in a flimsy T-shirt and shorts. When a finance minister says there will be 2.3 per cent growth in the economy next year it would still be helpful to know what the risk is of a recession.

Chapter 8 showed that, when produced by a computer, these single number forecasts are usually just an average of all the possibilities that might ensue. But often averages tell us nothing. Someone once argued that, if a person sits in their kitchen with one foot in the oven and the other in the freezer, on average they'll be comfortable. The average number drawn in Britain's National Lottery is thirty. It's a fact, but is that information helpful? Of course it's of no use at all. If I'm a baker and I'm told the computer has forecast that there'll be a demand for 428 cream cakes tomorrow, I'll have little idea of how many I should make if I'm more worried about disappointing customers than I am about being left with surplus cakes on my shelves. To make sensible decisions we need to be told what the other possibilities are, what chances they have of occurring and what the extreme outcomes look like. And, if the forecaster doesn't know, they should play it straight and tell us. The advice is writ large – be cautious about a single number, or a single event, prediction.

## Ineptitude

Ineptitude is perhaps an unkind word but it's appropriate when a forecaster does a worse job of prediction than they should. This can apply even if their intentions are honourable and even when they think they've done work of sterling quality. Unfortunately, ineptitude can be masked by consensus and confidence. If all forecasters agree that so-and-so will happen it's tempting to think they must be right. But what if they all read the same newspaper or listen to the same news channel? What if they all have the same background or specialism and have little idea what the world beyond looks like? Consensus is no guarantee of competence – in fact the opposite may be true. Only if consensus has been achieved through some carefully structured process, like the Delphi method or a supermeeting (see Chapter 5), can we relax a little and give the forecast some credence.

As to confidence, we've seen that unjustified confidence is the trademark of the media-loving pundit; more interested in booking their appearance on next week's programme than providing an unadulterated view of tomorrow. It's more likely to be a sign of a bad prediction. Think of Philip Tetlock's hedgehogs whom we met in Chapter 4. They were completely confident in their restricted ideas about the world, like people peering at the world through a small window, assured that there's nothing else to see.

So, if confidence and consensus can be misleading, what else can help us flush out the inept prediction that might cost us dearly? Much depends on whether it's based on human judgement or a computer. If it's judgement we need to know something about the judge. Have they had long-term experience in what they are forecasting and plenty of practice in turning their experience into forecasts? Have they had the opportunity to learn from regular, unambiguous feedback to tell them how reliable their forecasts are? Better still, if they can document the reasons for their forecasts then we have an audit trail. This should yield insights into the plausibility of their assumptions and how they've interpreted the available information. In Chapter 1 we saw that, under the right conditions, judgement can be brilliant. But we should take care if the forecaster's experience is scant, if they can't supply us with their track record and if their rationale bamboozles us with jargon or reveals an overly narrow view of the world. Be careful when a scientific-looking computer model is really an embodiment of someone's judgemental biases or tunnel vision. Recall from Chapter 4 Andy Haldane's crisis – hit economic forecasters at the Bank of England whose models assumed that people never behave irrationally. Remember also the pharmaceutical forecasters whom we met in Chapter 3, who cheerfully manipulated their model until it matched what they wanted it to say.

Beware also the colourful narrator who can pull a single story from memory, like a rabbit from a magician's hat, and base their whole forecast on that. 'Last time we tried this type of product promotion it was a huge success. I even got a hug from the boss ... we should sit back and watch our sales graph soar.' Or: 'the last physics graduate we hired was a dead loss. He spent most of his time offending everyone – including our best customers. So I don't have many hopes for the one they recruited last week.' It's shocking how selective, restricted and unreliable our memory can be – we can invent the past with the same unjustified confidence that we can presume to know the future. There are even stories of people who wrongly believed they were once lost in a shopping mall as a child, or who've been tricked into believing they took a ride in a hot air balloon.[8] When it comes to having an accurate view of what happened in the past, the electronic memory of the computer can trounce our grey cells. And it can recall as many cases as we like, rather than the one case that we might remember simply because it happened to be untypical.

Forecasters are usually too ready to replace the computer's forecast with their own. Too often they see its forecasts, rooted in past data, as irrelevant in a world of apparent continuous change. Each new twitch in a graph is seen to herald a fresh world disconnected from the past. Novelty can be interesting and exciting, but it doesn't mean that every new day is going to be fundamentally different. The forecaster who insists on discounting historic information altogether should tell us why. Historic information is likely to be more valuable than most planners recognise. And when there's plenty of it available, the computer won't only have a more reliable record of the past, it will usually outperform us in its search patterns and trends.

Even so, if these patterns and trends don't make sense, we are not obliged to unquestioningly accept them. They might be bringing us brilliant insights into a world we have long misunderstood. But they might also be spurious – freak patterns in our data, thrown up by an unthinking machine.

So we should be wary of the forecaster who operates at the extremes. Be suspicious if they spend their day chewing pens, or scratching their chins, trying to estimate what next week or next year will bring, while an expensive computer, replete with data, sits idly by. But be doubtful if they obediently accept the computer's surprising declarations, like the sat nav user heading for a non-existent bridge.

Perhaps it's time to design a wanted poster with the aim of stopping the bad prediction in its tracks. The headline will describe its transgressions as contamination, incompleteness or ineptitude, or some combination of these sins. Then the poster will need a paragraph to alert the public to the dangers of encountering it, while pointing out that it comes in many different guises. We'll warn that it can cloak itself in expertise, assurance, respectability or mathematical complexity; that it's likely to exploit our desire for reassurance and certainty; that it often pretends to be working in our interests, while serving someone else's ends. Most shocking of all, we'll need to say that it might be hiding in our own heads – waiting to deceive us as we plan our personal futures.

## A less-than-perfect prediction machine

Finally, what about the good prediction? In the last chapter I asked if you'd buy a perfect prediction machine from me. Fortunately, my machine exists only in the world of make-believe, sitting on an imaginary shelf alongside perpetual motion machines and vending machines that always deliver their goods when you've inserted your money. In the real world my machine could only give you its best assessments of probabilities for future events or, sometimes, an honest admission that it didn't know what the future holds. But at the very least it would alert you to the reality that the future is uncertain. Even the biggest decisions can be made ignoring this fact.

As the reverberations of Britain's shock exit from the EU continued, it emerged that Prime Minister David Cameron had banned his civil servants from making contingency plans in the event of a vote to leave. Buoyed by his surprise election victory a year earlier, it appears that he had regarded it as a certainty that the British public would vote to remain. When Deputy Prime Minister Nick Clegg challenged the decision to hold a referendum, he was 'breezily told that all would be well, of course it would be won'.[9]

There were only two possible outcomes of the referendum: leave or remain. There were simply no other unknown possibilities waiting to pounce out of the mist. My forecasting machine would have reminded David Cameron that both were possible; that he needed to plan for each eventuality. A good prediction should itself act like the perfect sceptic, alerting us to opportunities we may have too easily dismissed, but also challenging our comfortable complacency when needed.

And if it doesn't, be sceptical.

# NOTES

## Introduction

1  Harford, T., *Financial Times*, 9 August 2008.
2  Grim, R., Huffington Post, 5 July 2009.
3  Lawson, D., *Sunday Times*, 21 December 2014.

## Chapter 1: Neurons galore

1  www.bbc.co.uk/news/election-2015-32336071
2  www.standard.co.uk/news/uk/general-election-2015-britains-biggest-political-punter-loses-205500-on-hung-parliament-bet-after-winning-193000-on-scottish-referendum-wager-10237309.html
3  Worthen, B. (2003). 'Future results not guaranteed; contrary to what vendors tell you, computer systems alone are incapable of producing accurate forecasts.', *CIO*, 16 (19), p. 1.
4  Todorov, A., Mandisodza, A. N., Goren, A., & Hall, C. C. (2005). 'Inferences of competence from faces predict election outcomes', *Science*, 308 (5728), pp. 1623–6. Note that only the 2004 election involved true predictions. The results for the 2000 and 2002 elections were based on 'retrospective predictions'. Nevertheless there was no apparent difference in the accuracy of the two types of prediction.
5  Armstrong, J. S., Green, K. C., Jones, R. J., & Wright, M. J. (2010). 'Predicting elections from politicians' faces', *International Journal of Public Opinion Research*, 22 (4), pp. 511–22.
6  Sussman, A. B., Petkova, K., & Todorov, A. (2013). 'Competence ratings in US predict presidential election outcomes in Bulgaria', *Journal of Experimental Social Psychology*, 49 (4), pp. 771–5.
7  Ambady, N., & Rosenthal, R. (1993). 'Half a minute: Predicting teacher evaluations from thin slices of nonverbal behavior and physical attractiveness', *Journal of Personality and Social Psychology*, 64 (3), p. 431.

8     DeGroot, T., & Motowildo, S. J. (1999). 'Why visual and vocal interview cues can affect interviewers' judgments and predict job performance', *Journal of Applied Psychology*, 84 (6), pp. 986–93.

9     Hall, J. A., Roter, D. L., & Rand, C. S. (1981). 'Communication of effect between patient and physician', *Journal of Health and Social Behavior*, 22 (1), pp. 18–30.

10    Ambady, N., LaPlante, D., Nguyen, T., Rosenthal, R., & Levinson, W. (2002). 'Surgeon's tone of voice: A clue to malpractice history', *Surgery*, 132, pp. 5–9.

11    Chase, W. G. & Simon, H. A. (1973). 'The mind's eye in chess', in Chase, W. G. (ed.), *Visual Information Processing*, (London: Academic Press).

12    Goldstein, D. G., & Gigerenzer, G. (1999). 'The recognition heuristic: How ignorance makes us smart', in *Simple Heuristics That Make Us Smart*, (Oxford: Oxford University Press) pp. 37–58.

13    Ayton, P., Önkal, D., & McReynolds, L. (2011). 'Effects of ignorance and information on judgments and decisions', *Judgment and Decision Making*, 6 (5), pp. 381–91.

14    Scheibehenne, B., & Bröder, A. (2007). 'Predicting Wimbledon 2005 tennis results by mere player name recognition', *International Journal of Forecasting*, 23 (3), pp. 415–26.

15    Also see Serwe, S., & Frings, C. (2006). 'Who will win Wimbledon? The recognition heuristic in predicting sports events', *Journal of Behavioral Decision Making*, 19 (4), pp. 321–32. Recognition has also been used to successfully predict other sports events, such as the 2004 European Football Championship [see: Pachur, T., & Biele, G. (2007). 'Forecasting from ignorance: The use and usefulness of recognition in lay predictions of sports events', *Acta Psychologica*, 125 (1), pp. 99–116], or which of two Canadian hockey players has more career points [see: Snook, B., & Cullen, R. M. (2006). 'Recognizing National Hockey League greatness with an ignorance-based heuristic', *Canadian Journal of Experimental Psychology/Revue Canadienne de Psychologie Expérimentale*, 60 (1), p. 33].

16    Czerlinski J., Gigerenzer G. & Goldstein D. G. (1999). 'How good are simple heuristics?', in: Gigerenzer, G., Todd, P. M., the ABC Research Group (eds.) *Simple Heuristics That Make Us Smart*, (Oxford: Oxford University Press), pp. 97–118.

17    Note that while the computer gives you an exact forecast of a house's price, Take the Best can only be used to predict which of two houses will have the highest price.

18    Gigerenzer, G. (2011). 'What are natural frequencies?', *British Medical Journal*, 343, d6386.

19    Shedler, J. K., Jonides, J., & Manis, M. (November, 1985). 'Availability: Plausible but questionable', Paper presented at the 26th annual meeting of the Psychonomic Society, Boston, MA. Cited in Jonides, J., & Jones, C. M. (1992). 'Direct coding for frequency of occurrence', *Journal of Experimental Psychology: Learning, Memory, and Cognition*, 18(2), p. 368.

20    Tversky, A., & Kahneman, D. (1983). 'Extensional versus intuitive reasoning: the conjunction fallacy in probability judgment', *Psychological Review*, 90 (4), p. 293.

21    Gigerenzer G. (2000). *Adaptive Thinking: Rationality in the Real World*, (New York: Oxford University Press), Chapter 4.

22    Apostol, T. M. (1969). 'A short history of probability', in *Calculus, Volume II* (2nd edition). (Chichester: Wiley).

23    www.scientificamerican.com/article/why-does-the-brain-need-s/

24    For example, see Kahneman, D. (2011). *Thinking, Fast and Slow*, (London: Allen

Lane), p. 41. Though there is some controversy about this, see: www.scientifi-camerican.com/article/thinking-hard-calories/

25   'Fildes, R., & Goodwin, P. (2007). Against your better judgment? How organisations can improve their use of management judgment in forecasting', *Interfaces*, 37 (6), pp. 570–76.

## Chapter 2: A judgement on judgement

1   http://www.hscic.gov.uk/catalogue/PUB16988/obes-phys-acti-diet-eng-2015. pdfhttp://www.hscic.gov.uk/catalogue/PUB16988/obes-phys-acti-diet-eng-2015.pdf

2   Shermer, M. (2011). *The Believing Brain*, (Boston: Dutton).

3   The experiment is cited in Tetlock, P. (2005). *Expert Political Judgment: How good is it? How can we know?*, (Princeton NJ: Princeton University Press).

4   www.howtogetyourownway.com

5   Bogan, V., & Just, D. (2009). 'What drives merger decision making behavior? Don't seek, don't find, and don't change your mind', *Journal of Economic Behavior & Organization*, 72 (3), pp. 930–43.

6   Cipriano, M., & Gruca, T. S. (2015). 'The power of priors: how confirmation bias impacts market prices', *Journal of Prediction Markets*, 8, pp. 34–56.

7   Wagenaar, W. A. (1994). 'The subjective probability of guilt', in Wright, G. & Ayton. P. (eds.) *Subjective Probability*, (Chichester: Wiley).

8   Source: National Safety Council (USA).

9   Lichtenstein, S., Slovic, P., Fischhoff, B., Layman, M., & Combs, B. (1978). 'Judged frequency of lethal events', *Journal of Experimental Psychology: Human Learning and Memory*, 4, p. 551.

10  Johnson, E. J., Hershey, J., Meszaros, J., & Kunreuther, H. (1993). 'Framing, probability distortions, and insurance decisions', *Journal of Risk and Uncertainty*, 7, pp. 35–51.

11  Tversky, A., & Kahneman, D. (1974). 'Judgment under uncertainty: Heuristics and biases', *Science*, 185 (4157), pp. 1124–31.

12  Lawrence, M., Goodwin, P., O'Connor, M., & Önkal, D. (2006). 'Judgmental forecasting: A review of progress over the last 25 years', *International Journal of Forecasting*, 22 (3), pp. 493–518.

13  McKenzie, C. R., & Liersch, M. J. (2011). 'Misunderstanding savings growth: Implications for retirement savings behavior', *Journal of Marketing Research*, 48 (SPL), S1-S13.

14  Wagenaar, W. A., & Sagaria, S. D. (1975). 'Misperception of exponential growth', *Perception & Psychophysics*, 18 (6), pp. 416–22.

15  Wright, G. & Whalley, P. (1983). 'The super-additivity of subjective probability', in: Stigum, B. P., & Wenstop, F. (eds.) *Foundation of Utility and Risk Theory with Applications*, (Dordrecht: Reidel).

16  Ben-David, I., Graham, J. R., & Harvey, C. R. (2013). 'Managerial miscalibration', *Quarterly Journal of Economics*, 128, pp. 1547–84.

17  Soll, J. B., & Klayman, J. (2004). 'Overconfidence in interval estimates', *Journal of Experimental Psychology: Learning, Memory, and Cognition*, 30 (2), p. 299.

## Chapter 3: More bytes than we can chew

1   Cukier, K. N. & Mayer-Schoenberger, V. (2013). 'The rise of big data', *Foreign Affairs*, May/June issue.

2    Gage, D. (2014). 'Big Data Uncovers Some Weird Correlations', *Wall Street Journal*, 23 March.

3    Ibid.

4    Hays, C. (2004). 'What Wal-Mart knows about customers' habits', *New York Times*, 14 November.

5    Gage, D. (2014). 'Big Data Uncovers Some Weird Correlations', *Wall Street Journal*, 23 March.

6    Anderson, C. (2008). *WIRED*, 23 June 2008.

7    Quoted in Dormehl. L. (2014). *The Formula*, (London: WH Allen), p. 132.

8    Cahill, L. E., Chiuve, S. E., Mekary, R. A., Jensen, M. K., Flint, A. J., Hu, F. B., & Rimm, E. B. (2013). 'Prospective study of breakfast eating and incident coronary heart disease in a cohort of male US health professionals', *Circulation*, 128 (4), pp. 337–43.

9    Lovelock, J. (2014). *A Rough Ride to the Future*, (London: Penguin), p. 56.

10   Quinn, G. E., Shin, C. H., Maguire, M. G., & Stone, R. A. (1999). 'Myopia and ambient lighting at night', *Nature*, 399 (6732), pp. 113–14.

11   Gwiazda, J., Ong, E., Held, R., & Thorn, F. (2000). 'Vision: Myopia and ambient night-time lighting', *Nature*, 404, p. 144.

12   Quoted in: Sedensky, M. (2016), 'Trump's surprise victory proves that big data is "still in its infancy"', Businessinsider.com, 11 November, 2016.

13   Cave, K.(2016). 'Trump victory: Slap in the face for big data or desperate cry for more?', Idgconnect.co, 9 November, 2016.

14   Timms, A. (2016). 'Is Donald Trump's surprise win a failure of big data? Not really.' *Fortune*, 14 November, 2016.

15   Alan Lichtman's 'keys method' called the 2016 election correctly and has had an excellent track record in predicting presidential elections since 1984. The method is not based on big data; instead, it uses an assessment of whether thirteen statements are true or false. For example: 'The challenging party candidate is not charismatic or a national hero'. When six or more statements are judged to be false, the candidate of the party not currently in office is predicted to win.

16   Prokop, A. (2016). 'Nate Silver's model gives Trump an unusually high chance of winning. Could he be right?' vox.com, 3 November, 2016.

17   Fildes, R., Goodwin, P., Lawrence, M., & Nikolopoulos, K. (2009). 'Effective forecasting and judgmental adjustments: an empirical evaluation and strategies for improvement in supply-chain planning', *International Journal of Forecasting*, 25, pp. 3–23.

18   Goodwin, P., Önkal, D., & Lawrence, M. (2011). 'Improving the role of judgment in economic forecasting', in Clements, M. P., & Hendry, D. F. (eds.) *The Oxford Handbook of Economic Forecasting*, (Oxford: Oxford University Press).

19   http://uk.reuters.com/article/2006/12/22/uk-germany-satnav-idUKL2284422820061222

20   Fildes, R., Goodwin, P., Lawrence, M., & Nikolopoulos, K. (2009). 'Effective forecasting and judgemental adjustments: an empircal evaluation and strategies for improvement in supply-chain planning'. *International Journal of Forecasting*, 25, pp. 3–23.

21   Carr, N. (2015).*The Glass Cage, Where Automation is Taking Us*, (London: Penguin).

22   Davis, F. D., Lohse, G. L., & Kottemann, J. E. (1994). 'Harmful effects of seemingly

helpful information on forecasts of stock earnings', *Journal of Economic Psychology*, 15, pp. 253–67.

23  Arkes, H. R., Dawes, R. M., & Christensen, C. (1986). 'Factors influencing the use of a decision rule in a probabilistic task', *Organizational Behavior and Human Decision Processes*, 37, pp. 93–110.

24  Lim, J. S., & O'Connor, M. (1995). 'Judgemental adjustment of initial forecasts: Its effectiveness and biases', *Journal of Behavioral Decision Making*, 8, pp. 149–68.

25  Taleb, N. N. (2005). *Fooled by Randomness: The Hidden Role of Chance in Life and in the Markets*, (New York: Random House).

26  Önkal, D., Goodwin, P., Thomson, M., Gönül, S., & Pollock, A. (2009). 'The relative influence of advice from human experts and statistical methods on forecast adjustments', *Journal of Behavioral Decision Making*, 22, pp. 390–409.

27  Fischhoff, B., & Beyth, R. (1975). 'I knew it would happen: Remembered probabilities of once future things', *Organizational Behavior and Human Performance*, 13, pp. 1–16.

28  Cassar, G., & Craig, J. (2009). 'An investigation of hindsight bias in nascent venture activity', *Journal of Business Venturing*, 24, pp. 149–64.

29  Biais, B., & Weber, M. (2009). 'Hindsight bias, risk perception, and investment performance', *Management Science*, 55, pp. 1018–29.

30  Fildes, R., & Stekler, H. (2002). 'The state of macroeconomic forecasting', *Journal of Macroeconomics*, 24, pp. 435–68.

31  Goodwin, P. (2009). 'New evidence on the value of combining forecasts', *Foresight: The International Journal of Applied Forecasting*, 12, pp. 33–5.

32  Graefe, A., Armstrong, J. S., Jones, R. J., & Cuzán A. G. (2014). 'Combining forecasts: An application to elections', *International Journal of Forecasting*, 30, pp. 43–54.

**Chapter 4: Spot the expert**

1  Shanteau, J. (1992). 'The psychology of experts an alternative view', in Wright. G, & Bolger, F. (eds.) *Expertise and Decision Support*, (New York: Plenum), pp. 11–23.

2  Tetlock, P. E. (2005). *Expert Political Judgment*, (Princeton: Princeton University Press).

3  Denrell, J., & Fang, C. (2010). 'Predicting the next big thing: success as a signal of poor judgment', *Management Science*, 56, pp. 1653–67.

4  Powdthavee, N., & Riyanto, Y. E. (2015). 'Would you pay for transparently useless advice? A test of boundaries of beliefs in the folly of predictions', *Review of Economics and Statistics*, 97, pp. 257–72. The researchers used a method demonstrated by the magician Derren Brown on British television. In his show *The System* he convinced a woman that he was able to predict six consecutive wins at a horse race. In fact, for the first race, several of a large number of people had received different predictions so that all possible winners were covered. Brown then contacted those who had being given a correct prediction and repeated the process for the next race. Gradually the number of participants with all correct predictions was winnowed down until one person – who had received six out of six correct forecasts – was left. From her perspective the predictions must have seemed to have been miraculously accurate.

5  Armstrong, J. S. (1980). 'The seer-sucker theory: The value of experts in forecasting', *Technology Review*, June/July, 16–24.

6    Cocozza, J. J., & Steadman, H. J. (1978). 'Prediction in psychiatry: an example of misplaced confidence in experts', *Social Problems*, 25, pp. 265–76.

7    Hagen, M. (1997). *Whores of the Court: The Fraud of Psychiatric Testimony and the Rape of American Justice*, (New York: Harper Collins).

8    Torngren, G., & Montgomery, H. (2004). 'Worse than chance? Performance and confidence among professionals and laypeople in the stock market', *Journal of Behavioral Finance*, 5, pp. 148–53.

9    Cowles 3rd, A. (1933). 'Can stock market forecasters forecast?' *Econometrica*, 1, pp. 309–24.

10   Tetlock, P. E. (2005). *Expert Political Judgment*, (Princeton: Princeton University Press), p. 2.

11   Lipinski, M., Froelicher, V., Atwood, E., Tseitlin, A., Franklin, B., Osterberg, L., Do, D., & Myers, J. (2002). 'Comparison of treadmill scores with physician estimates of diagnosis and prognosis in patients with coronary artery disease', *American Heart Journal*, 143, pp. 650–58.

12   Berner, E. S., & Graber, M. L. (2008). 'On overconfidence and diagnostic error reply', *American Journal of Medicine*, 121, p. E19.

13   Fildes, R., Goodwin, P., Lawrence, M., & Nikolopoulos, K. (2009). 'Effective forecasting and judgmental adjustments: an empirical evaluation and strategies for improvement in supply-chain planning', *International Journal of Forecasting*, 25, pp. 3–23.

**Chapter 5: Group power**

1    Asch, S. E. (1951). 'Effects of group pressure upon the modification and distortion of judgment', in Guetzkow, H. (ed.) *Groups, Leadership and Men*, (Pittsburgh, PA: Carnegie Press).

2    Janis, I. L. (1972). *Victims of Groupthink*, (New York: Houghton Mifflin).

3    King, A. & Crewe, I. (2003). *The Blunders of our Governments*, (Oxford: Oneworld Publications).

4    Heath, A. (2015). *Daily Telegraph*, 24 September 2015.

5    Hermann, A., & Rammal, H. G. (2010). 'The grounding of the "flying bank"', *Management Decision*, 48, pp. 1048–62.

6    Eaton, J. (2001). 'Management communication: the threat of groupthink', *Corporate Communications: An International Journal*, 6, pp. 183–92.

7    Sunstein, C. R., & Hastie, R. (2014). 'Making Dumb Groups Smarter', *Harvard Business Review*, 92, pp. 90–98.

8    Morgan, G. (1986). *Images of Organization*, (Beverly Hills, CA: Sage). Another alleged example of groupthink is NASA's disastrous decision to launch the Challenger space shuttle in 1997. This is widely discussed in other books.

9    Cole, D. (2015), *Alaska Dispatch News*, 24 April 2015.

10   Hueffer, K., Fonseca, M. A., Leiserowitz, A., & Taylor, K. M. (2013). 'The wisdom of crowds: Predicting a weather and climate-related event', *Judgment and Decision Making*, 8, pp. 91–105.

11   Surowiecki, J. (2004). *The Wisdom of Crowds*, (London: Little Brown).

12   Galton, F. (1907). 'Vox populi'. *Nature*, 75, pp. 450–51.

13   Lorenz, J., Rauhut, H., Schweitzer, F., & Helbing, D. (2011). 'How social influence can undermine the wisdom of crowd effect', *Proceedings of the National Academy of Sciences*, 108, pp. 9020–25.

14     Ashton, A. H., & Ashton, R. H. (1985). 'Aggregating subjective forecasts: Some empirical results', *Management Science*, 31, pp. 1499–508.

15     Davis-Stober, C. P., Budescu, D. V., Dana, J., & Broomell, S. B. (2014). 'When is a crowd wise?' *Decision*, 1, p. 79.

16     Armstrong, J. S. (2006). 'Should the forecasting process eliminate face-to-face meetings?', *Foresight: The International Journal of Applied Forecasting*, Fall, 3–8.

17     Maier, N. R. F. (1963). 'Problem Solving Discussions and Conferences', (New York: McGraw Hill).

18     Sunstein, C. R., & Hastie, R. (2014). 'Making dumb groups smarter', *Harvard Business Review*, 92, pp. 90–98.

19     Atanasov, P. D., Rescober, R., Stone, E., Swift, S. A., Servan-Schreiber, E., Tetlock, P. E., Ungar, L., & Mellers, B. (2016). 'Distilling the wisdom of crowds: Prediction markets versus prediction polls', *Management Science*, Forthcoming.

20     Panels of these sizes are recommended by Rowe, G., & Wright, G. (2001). 'Expert opinions in forecasting: The role of the Delphi technique', in Armstrong, J. S. (ed.) *Principles of Forecasting*, (Boston: Kluwer Academic Publishers).

21     Ibid.

22     Bolger, F., Stranieri, A., Wright, G., & Yearwood, J. (2011). 'Does the Delphi process lead to increased accuracy in group-based judgmental forecasts or does it simply induce consensus amongst judgmental forecasters?', *Technological Forecasting and Social Change*, 78 (9), pp. 1671–80.

23     www.the-numbers.com/movie

24     Pennock, D. M., Lawrence S., Giles C. L., & Nielsen, F. A. (2001). 'The Real Power of Artificial Markets', *Science*, 291, pp. 987–8.

25     Chen, K-Y., & Plott, C. (2002). 'Information aggregation mechanisms: concept, design and implementation for a sales forecast problem', *Caltech Social Science Working Paper No. 1331*.

26     Gurkaynak, R., & Wolfers, J. (2006). 'Macroeconomic derivatives: An initial analysis of market-based macro forecasts, uncertainty, and risk', (No. Working Paper: 11929), National Bureau of Economic Research.

### Chapter 6: Forecasting ourselves

1     We define unrealistic optimism as a tendency to overestimate the true probability of good events occurring and underestimate the probability of negative events happening.

2     Stavrova, O., & Ehlebracht, D. (2015). 'Cynical Beliefs About Human Nature and Income: Longitudinal and Cross-Cultural Analyses', *Journal of Personality and Social Psychology*, Forthcoming.

3     Taylor, S. E., & Brown, J. D. (1994). 'Positive illusions and well-being revisited: separating fact from fiction', *Psychological Bulletin*, 116, pp. 21–7.

4     Hevey, D., McGee, H. M., & Horgan, J. H. (2014). 'Comparative optimism among patients with coronary heart disease (CHD) is associated with fewer adverse clinical events 12 months later', *Journal of Behavioral Medicine*, 37, pp. 300–307.

5     Sharot, T. (2011). *The Optimism Bias: A Tour of the Irrationally Positive Brain*, (New York: Vintage).

6     Alloy, L. B., & Abramson, L. Y. (1979). 'Judgment of contingency in depressed and non-depressed students: Sadder but wiser?', *Journal of Experimental Psychology: General*, 108, p. 441.

7    Ackermann, R., & DeRubeis, R. J. (1991). 'Is depressive realism real?', *Clinical Psychology Review*, 11 (5), pp. 565–84.

8    Moore, M. T., & Fresco, D. M. (2012). 'Depressive realism: a meta-analytic review', *Clinical Psychology Review*, 32(6), pp. 496–509.

9    Varki, A. (2009). 'Human uniqueness and the denial of death', *Nature*, 460, pp. 684–84. Also see: Sharot, T. (2011). *The Optimism Bias: A Tour of the Irrationally Positive Brain*, (New York: Vintage).

10   Weinstein, N. D. (2001). 'Smokers' recognition of their vulnerability to harm', in Slovic, P. (ed.) *Smoking Risk, Perception, and Policy*, (Thousand Oaks, CA: Sage), pp. 81–96.

11   Kim, H. K., & Niederdeppe, J. (2013). 'Exploring optimistic bias and the integrative model of behavioral prediction in the context of a campus influenza outbreak', *Journal of Health Communication*, 18, pp. 206–22.

12   Morwitz, V. G., & Fitzsimons, G. J. (2004). 'The mere-measurement effect: Why does measuring intentions change actual behavior?', *Journal of Consumer Psychology*, 14, pp. 64–74.

13   Weinstein, N. D. (1980). 'Unrealistic optimism about future life events', *Journal of Personality and Social Psychology*, 39, p. 806.

14   Lewis, C. S. (1952). *Mere Christianity*, (New York: Macmillan).

15   Epley, N., & Dunning, D. (2000). 'Feeling "holier than thou": are self-serving assessments produced by errors in self or social prediction?', *Journal of Personality and Social Psychology*, 79, p. 861.

16   Perloff, L. S., & Fetzer, B. K. (1986). 'Self–other judgments and perceived vulnerability to victimization', *Journal of Personality and Social Psychology*, 50, p. 502.

17   Klein, C. T., & Helweg-Larsen, M. (2002). 'Perceived control and the optimistic bias: A meta-analytic review', *Psychology and Health*, 17, pp. 437–46.

18   Norem, J. K. (2001). 'Defensive pessimism, optimism, and pessimism', in Chang, E. (ed.), *Optimism and Pessimism: Implications for Theory, Research, and Practice*, (Washington, DC: American Psychological Association), pp. 77–100.

19   Norem, J. K., & Burdzovic Andreas, J. (2006). 'Understanding journeys: Individual growth analysis as a tool for studying individual differences in change over time', in Ong, A. D., & van Dulmen, M. (eds.), *Handbook of Methods in Positive Psychology*, (London: Oxford University Press) pp. 1036–58.

20   Norem, J. (2001). *The Positive Power of Negative Thinking*, (Cambridge, MA: Basic Books).

21   Norem. J. K., & Cantor, N. (1986). 'Defensive pessimism: Harnessing anxiety as motivation', *Journal of Personality and Psychology*, 51, pp. 1208–17.

22   Stewart, A. (1967). *Bedsitter Images*, CBS Records.

23   Wilson, T. D., & Gilbert, D. T. (2005). 'Affective forecasting: knowing what to want', *Current Directions in Psychological Science*, 14, pp. 131–34.

24   Schkade, D. A., & Kahneman, D. (1998). 'Does living in California make people happy? A focusing illusion in judgements of life satisfaction', *Psychological Science*, 9, pp. 340–46.

25   Campbell, W. K., & Sedikides, C. (1999). 'Self-threat magnifies the self-serving bias: A meta-analytic integration', *Review of General Psychology*, 3, p. 23.

26   Gilbert, D. T., Pinel, E. C., Wilson, T. D., Blumberg, S. J., & Wheatley, T. P. (1998). 'Immune neglect: a source of durability bias in affective forecasting', *Journal of Personality and Social Psychology*, 75, p. 617.

27   Suh, E., Diener, E., & Fujita, F. (1996). 'Events and subjective well-being: only recent events matter', *Journal of Personality and Social Psychology*, 70, p. 1091.

28  Gilbert, D. T., Pinel, E. C., Wilson, T. D., Blumberg, S. J., & Wheatley, T. P. (1998). 'Immune neglect: a source of durability bias in affective forecasting', *Journal of Personality and Social Psychology*, 75, p. 617.

29  Wang, J., Novemsky, N., & Dhar, R. (2007). 'How predictions differ from actual adaptation to durable products', *Advances in Consumer Research*, 34, pp. 547–8.

30  Ibid.

31  Kruger, J., & Dunning, D. (1999). 'Unskilled and unaware of it: How difficulties in recognizing one's own competence lead to inflated self-assessments', *Journal of Personality and Social Psychology*, 77, pp. 1121–34.

32  McCormick, I. A., Walkey, F. H., & Green, D. E. (1986). 'Comparative perceptions of driver ability—a confirmation and expansion', *Accident Analysis & Prevention*, 18, pp. 205–8. Though 80 per cent of drivers being above average is not necessarily a statistical impossibility. It depends on how you define 'average'. If you scored drivers for their ability and used the mean score as the average, some extremely poor drivers could make this possible. For example, if the scores out of 100 for five drivers were 2, 70, 85, 90, 98 the average would be 69, and 80 per cent of the drivers would exceed this score. If instead you use the median as the average, then only 50 per cent of drivers could be above average.

33  Felson, R. B. (1981). 'Ambiguity and bias in the self-concept', *Social Psychology Quarterly*, 44, pp. 64–9. Also see Larwood, L., & Whittaker, W. (1977). 'Managerial myopia: self-serving biases in organizational planning', *Journal of Applied Psychology*, 62, pp. 194–8.

34  Meeran, S., Goodwin, P., & Yalabik, B. (2016). 'A parsimonious explanation of observed biases when forecasting one's own performance', *International Journal of Forecasting*, 32, pp. 112–20.

**Chapter 7: I spy impostors!**

1  Ross, S. A. (1973). 'On the economic theory of agency: The principal's problem', *American Economic Review*, 63, pp. 134–39. Also see Mitnick, B. M. (1974). 'The Theory of Agency: The Concept of Fiduciary Rationality and Some consequences', unpublished PhD Dissertation. Department of Political Science, University of Pennsylvania. Univ. Microfilms No. 74–22, p. 881.

2  Goodwin, P. (1998). 'Enhancing judgmental sales forecasting: the role of laboratory research', in Wright, G., & Goodwin, P. (eds.) *Forecasting with Judgment*, (Chichester: Wiley) pp. 91–112. Also see Goodwin, P. (2016). 'Misbehaving Aagents', *Foresight: The International Journal of Applied Forecasting*. Some of the discussion here is based closely on this article.

3  Kirchgässner, G., & Müller, U. K. (2006). 'Are forecasters reluctant to revise their predictions? Some German evidence', *Journal of Forecasting*, 25 (6), pp. 401–13.

4  Nordhaus, W. D. (1987). 'Forecasting efficiency: concepts and applications', *Review of Economics and Statistics*, November, pp. 667–74.

5  Keren, G. (1997). 'On the calibration of probability judgments: Some critical comments and alternative perspectives', *Journal of Behavioral Decision Making*, 10 (3), pp. 269–78.

6  Price, P. C., & Stone, E. R. (2004). 'Intuitive evaluation of likelihood judgment producers: Evidence for a confidence heuristic', *Journal of Behavioral Decision Making*, 17 (1), pp. 39–57.

7  I have used the term forecasts here and in all that follows in this chapter, because

this is how these estimates are labelled and I did not want to overuse inverted commas. However, all of these so-called forecasts are really decisions.

8    Ottaviani, M., & Sørensen, P. N. (2006). 'The strategy of professional forecasting', *Journal of Financial Economics*, 81 (2), pp. 441–66. Also see Laster, D., Bennett, P. B., & Geoum, I. S. (1999). 'Rational bias in macroeconomic forecasts', *Quarterly Journal of Economics*, 114, pp. 293–318.

9    Henry, G. B., (1989). 'Wall Street economists: are they worth their salt?', *Business Economics*, 10, pp. 44–8.

10   Pierdzioch, C., Rülke, J. C., & Stadtmann, G. (2012). 'A Note on Forecasting Emerging Market Exchange Rates: Evidence of Anti-herding', *Review of International Economics*, 20 (5), pp. 974–84.

11   Pierdzioch, C., Rülke, J. C., & Stadtmann, G. (2010). 'New evidence of anti-herding of oil-price forecasters', *Energy Economics*, 32 (6), pp. 1456–9.

12   Pierdzioch, C., Rülke, J. C., & Stadtmann, G. (2013). 'Forecasting metal prices: Do forecasters herd?', *Journal of Banking & Finance*, 37 (1), pp. 150–58.

13   Batchelor, R. A., & Dua, P. (1990). 'Product differentiation in the economic forecasting industry', *International Journal of Forecasting*, 6, pp. 311–16.

14   Ashiya, M., & Doi, T. (2001). 'Herd behavior of Japanese economists', *Journal of Economic Behavior & Organization*, 46 (3), pp. 343–6.

15   Lamont, O. A. (2002). 'Macroeconomic forecasts and microeconomic forecasters', *Journal of Economic Behavior & Organization*, 48 (3), pp. 265–80.

16   Ibid.

17   http://www.valuewalk.com/2012/12/gary-shilling-makes-bold-2013-prediction-after-disastrous-2012-call/

18   Quoted in Lamont, O. A. (2002). 'Macroeconomic forecasts and microeconomic forecasters', *Journal of Economic Behavior & Organization*, 48(3), pp. 265–80.

19   www.metoffice.gov.uk/climate/uk/summaries/2014/summer

20   George Monbiot's blog for *The Guardian*, 15 June 2012.

21   Ibid.

22   Keynes, J. M. (1936). *General Theory of Employment, Interest and Money*, (London: Macmillan).

23   Hong, H., Kubik, J. D., & Solomon, A. (2000). 'Security analysts' career concerns and herding of earnings forecasts', *Rand Journal of Economics*, 31 (1), pp. 121–44.

24   Bewley, R., & Fiebig, D. G. (2002). 'On the herding instinct of interest rate forecasters', *Empirical Economics*, 27 (3), pp. 403–25.

25   Clements, M. P. (2014). 'Do US Macroeconomic Forecasters Exaggerate Their Differences?', Available at Social Science Research Network. SRN Paper no. 2496433.

26   Fildes, R., & Goodwin, P. (2007). 'Against your better judgment? How organizations can improve their use of management judgment in forecasting', *Interfaces*, 37 (6), pp. 570–76.

27   Fildes, R., & Hastings, R. (1994). 'The organization and improvement of market forecasting', *Journal of the Operational Research Society*, 45 (1), pp. 1–16.

28   Galbraith, C. S., & Merrill, G. B. (1996). 'The Politics of Forecasting: Managing the Truth', *California Management Review*, 38(2), pp. 29–43.

29   Bretschneider, S., & Gorr, W. (1992). 'Economic, organizational, and political influences on biases in forecasting state sales tax receipts', *International Journal of Forecasting*, 7 (4), pp. 457–66.

30   Deschamps, E. (2004). 'The impact of institutional change on forecast accuracy:

A case study of budget forecasting in Washington State', *International Journal of Forecasting*, 20 (4), pp. 647–57.

31  Dreher, A., Marchesi, S., & Vreeland, J. R. (2008). 'The political economy of IMF forecasts', *Public Choice*, 137 (1–2), pp. 145–71. Similar results were found in Aldenhoff, F. O. (2007). 'Are economic forecasts of the International Monetary Fund politically biased? A public choice analysis', *Review of International Organizations*, 2 (3), pp. 239–60.

32  www.bbc.co.uk/news/health Report on National Obesity Forum report 13 January 2014

33  www.bbc.co.uk/news/science-environment-17488450

34  www.jrf.org.uk/media-centre/uk-poverty-levels-forecast-rise-19562

35  IPCC Fifth Assessment Report, 2013, Chapter 12.

36  Lynas, M., 'Climate change explained – the impact of temperature rises', *The Guardian*, 14 April 2009.

37  Fildes, R., & Kourentzes, N. (2011). 'Validation and forecasting accuracy in models of climate change', *International Journal of Forecasting*, 27 (4), pp. 968–95.

38  www.theclimatebet.com

39  www.scientificamerican.com/article/climate-scientists-helped-create-a-spurious-pause-in-global-warming/

40  For example, Bell, L. (2015). *Scared Witless: Prophets and Profits of Climate Doom*, (Mt Vernon, WA: Stairway Press).

41  www.nytimes.com/2015/02/22/us/ties-to-corporate-cash-for-climate-change-researcher-Wei-Hock-Soon.html?_r=0.

42  Green, K. C., Armstrong, J. S., & Soon, W. (2009). 'Validity of climate change forecasting for public policy decision making', *International Journal of Forecasting*, 25 (4), pp. 826–32.

43  www.washingtontimes.com/news/2015/mar/24/j-scott-armstrong-missing-the-mark-on-climate-change/

## Chapter 8: The lure of one number

1  Bank of England. The forecast is reproduced with permission.

2  Miller, G. A. (1956). 'The magical number seven, plus or minus two: Some limits on our capacity for processing information', *Psychological Review*, 63, pp. 81–97.

3  Tversky, A. (1969). 'Intransitivity of Preferences', *Psychological Review*, 76, pp. 31–48.

4  *Prospect*, March 2015, p. 16.

5  Skidelsky, R. (2010). *Keynes: A Very Short Introduction*, (Oxford: Oxford University Press).

6  www.lottery.co.uk/statistics. Note that until October 2015 the UK lottery involved selecting numbers between 1 and 49, rather than 1 to 59.

7  Gigerenzer, G., Hertwig, R., van den Broek, E., Fasolo, B., Katsikopoulos, K. V. (2005). '"A 30 Per Cent Chance of Rain Tomorrow": How Does the Public Understand Probabilistic Weather Forecasts?', *Risk Analysis*, 25, pp. 623–9. Another interpretation provided by the authors is that, when the weather conditions are like those we are seeing today (e.g. a depression crossing from the Atlantic), in three out of ten cases there will be at least a trace of rain the next day.

8  Soyer, E., & Hogarth, R. M. (2012). 'The illusion of predictability: How regression statistics mislead experts', *International Journal of Forecasting*, 28 (3), pp. 695–711.

9   *Daily Telegraph*, 21 July 2014.

10  *The Independent*, 29 March 2013.

11  Ramos, M. H., van Andel, S. J., & Pappenberger, F. (2013). 'Do probabilistic forecasts lead to better decisions?', *Hydrology and Earth Systems Sciences*, 17, pp. 2219–32. Savelli, S., & Joslyn, S. (2013). 'The advantages of predictive interval forecasts for non-expert users and the impact of visualizations', *Applied Cognitive Psychology*, 27, pp. 527–41.

12  Yaniv, I., & Foster, D. P. (1995). 'Graininess of judgment under uncertainty: An accuracy-informativeness trade-off', *Journal of Experimental Psychology: General*, 124 (4), p. 424.

13  Du, N., Budescu, D. V., Shelly, M. K., & Omer, T. C. (2011). 'The appeal of vague financial forecasts', *Organizational Behavior and Human Decision Processes*, 114 (2), pp. 179–89.

14  Dalrymple, D. J. (1987). 'Sales forecasting practices: Results from a United States survey', *International Journal of Forecasting*, 3, pp. 379–91.

15  Klassen, R. D., & Flores, B. E. (2001). 'Forecasting practices of Canadian firms: Survey results and comparisons', *International Journal of Production Economics*, 70 (2), pp. 163–74.

16  Armstrong, J. S. (ed.) (2001). *Principles of Forecasting*, (Boston: Kluwer Academic Publishing).

17  Goodwin, P., & Wright, G. (2010). 'The limits of forecasting methods in anticipating rare events', *Technological Forecasting and Social Change*, 77, pp. 355–68.

18  Taleb, N. N. (2007). *The Black Swan*, (London: Allen Lane).

19  Stewart, I. 'The mathematical equation that caused the banks to crash', *The Observer*, 12 February 2012.

20  Teigen, K. H., Juanchich, M., & Riege, A. H. (2013). 'Improbable outcomes: Infrequent or extraordinary?', *Cognition*, 127 (1), pp. 119–39.

21  Beyth-Marom, R. (1982). 'How probable is probable? A numerical translation of verbal probability expressions', *Journal of Forecasting*, 1, pp. 257–69.

22  Teigen, K. H., Juanchich, M., & Riege, A. H. (2013). 'Improbable outcomes: infrequent or extraordinary?', *Cognition*, 127, pp. 119–39.

23  Budescu, D. V., Por, H-H., & Broomwell, S. B. (2012). 'Effective communication of uncertainty in the IPCC reports', *Climatic Change*, 113, pp. 181–200.

24  Hogarth, R. M., & Soyer, E. (2015). 'Communicating forecasts: The simplicity of simulated experience', *Journal of Business Research*, 68, pp. 1800–809.

**Chapter 9: Those who neglect history**

1   http://www.businessinsider.com/7-weirdest-things-you-can-bet-on-2013-7#ixzz3SecHRCXz

2   http://www.bhic.co.uk/facts-and-figures.html

3   http://www.bbc.co.uk/sport/0/football/24354124

4   Lichtman, A. J. (2006). 'The keys to the White House: forecast for 2008', *Foresight: The International Journal of Applied Forecasting*, Issue 3, pp. 5–9.

5   Green, K. C., & Armstrong, J. S. (2007). 'Structured analogies for forecasting', *International Journal of Forecasting*, 23, pp. 365–76.

6   Kahneman, D., & Lovallo, D. (1993). 'Timid choices and bold forecasts: A cognitive perspective on risk taking', *Management Science*, 39, pp. 17–31.

7   Office for National Statistics.

8   Forest Institute of Professional Psychology, Springfield.

9    Cassar, G. (2010). 'Are individuals entering self-employment overly optimistic? An empirical test of plans and projections on nascent entrepreneur expectations', *Strategic Management Journal*, 31, pp. 822–40.

10   *The Scotsman*, 27 August 2003.

11   Flyvbjerg, B. (2008). 'Curbing optimism bias and strategic misrepresentation in planning: Reference class forecasting in practice', *European Planning Studies*, 16 (1), pp. 3–21.

12   *West Briton*, 24 April 2015.

13   *Manchester Evening News*, 30 June 2005.

14   http://www.psychologicalscience.org/index.php/uncategorized/what-was-i-thinking-kahneman-explains-how-intuition-leads-us-astray.html

15   McNally, D. (2011). *Global Slump: The Economics and Politics of Crisis and Resistance*, (Oakland, CA: PM Press), p. 111.

16   This is based on the Case-Shiller US National Home Price Index.

17   UK Department for Transport.

18   This has been generated by a computer using a widely applied forecasting technique called Holt's method. If more data had been available, the damped-Holt's method could have been used as an alternative. This allows a damped trend, rather than a linear trend, to be projected into the future so that a slowing down of the trend can be taken into account in the forecasts.

19   Runkle, D. E. (1998). 'Revisionist history: How data revisions distort economic policy research', *Federal Reserve Bank of Minneapolis Quarterly Review*, 22, pp. 3–12.

20   Zarnowitz, V. (1967). 'An appraisal of short-term economic forecasts', *NBER occasional paper 104*, National Bureau of Economic Research, New York.

21   I retrospectively applied a model discussed in Lewis-Beck, M. S., Nadeau, R., & Bélanger, E. (2004). 'General election forecasts in the United Kingdom: a political economy model', *Electoral Studies*, 23, pp. 279–90. The model is: Vote share of governing party (per cent) = 42.7 – 0.97 x Inflation Rate + 0.27 Public approval of government record – 3.1 x No. of terms governing party has been in office + Effect of random factors. The formula was estimated by analysing twelve consecutive UK general elections, up to and including the 1997 election. The authors provided only a point forecast. The probabilities were calculated by me using standard assumptions implied by the authors' analysis e.g. that the noise was normally distributed.

22   Norris, P. (2001). 'Apathetic landslide: the 2001 British general election', *Parliamentary Affairs*, 54, pp. 565–89.

23   Whitely, P., 'Ballot blues', *The Guardian*, 1 March 2001.

24   http://news.bbc.co.uk/1/hi/uk/4507448.stm

25   Dormehl, L. (2014). *The Formula*, (London: WH Allen).

26   Ramsey, J. B., & Harris, R. (1977). 'Economic Forecasting – Models or Markets?: An Introduction to the Role of Econometrics in Economic Policy'. Institute of Economic Affairs, p. 86.

27   O'Brien, F. A. (2004). 'Scenario planning: lessons for practice from teaching and learning', *European Journal of Operational Research*, 152 (3), pp. 709–22.

28   Popper, K. (1957). *The Poverty of Historicism*, (London: Routledge and Kegan Paul).

29   Markopolos, H. (2010). *No One Would Listen: A True Financial Thriller*, (New York: Wiley).

30 Bazerman, M. H. & Watkins, M. D. (2008). *Predictable Surprises*, (Cambridge, MA: Harvard Business School Press).

31 Bell, D. (1979). 'The Information Society', in Dertouzos, M. L., & Moses, J. (eds.) *The Computer Age: A Twenty-year View*. (Cambridge, MA: The Massachusetts Institute of Technology Press), p. 195.

### Chapter 10: You can't tell me my forecast was wrong

1 US Department of Health and Human Services. Anthropometric reference data for children and adults: United States 2007–2010. Vital Health Statistics, Series 11, no. 252.

2 Lotto changed in October 2015. People now pick six numbers from fifty-nine (rather than forty-nine).

3 Goodwin, P. (2011). 'High on complexity, low on evidence: are advanced forecasting methods always as good as they seem?', *Foresight: the International Journal of Applied Forecasting*, 23, Fall, 10–12.

4 Zhu, S., Wang, J., Zhao, W., & Wang, J. (2011). 'A seasonal hybrid procedure for electricity demand forecasting in China', *Applied Energy*, 88, pp. 3807–15.

5 Gilliland, M., & Sglavo, U. (2010). 'Focus on forecasting: worst practices in business forecasting', *Analytics*, July/August.

6 Murphy, A. H., & Winkler, B. L. (1974). 'Subjective probability forecasting experiments in meteorology: Some preliminary results', *Bulletin of the American Meteorological Society*, 55, pp. 1206–16. Also see Stewart, T. R., Roebber, P. J., & Bosart, L. F. (1999). 'The importance of the task in analyzing expert judgment', *Organizational Behavior and Human Decision Processes*, 69, pp. 205–19.

7 http://www.theguardian.com/uk-news/the-northerner/2014/jan/30/how-often-does-it-rain-in-manchester

8 For example, a study by James Taylor was able to use six years' worth of half-hourly electricity data – that's 105,168 observations. See Taylor, J. W. (2010). 'Triple seasonal methods for short-term electricity demand forecasting', *European Journal of Operational Research*, 204 (1), pp. 139–52.

9 Gardner, D. (2010*). Future Babble*, (Toronto: McClelland and Stewart Ltd), p. 252.

10 'Global Trends 2015: A Dialogue About the Future With Nongovernment Experts', NIC 2000–02. The predictions in this document do not contain probabilities but many do acknowledge that there is uncertainty associated with them.

11 *Daily Mail*, 5 July 2015.

12 Makridakis, S., Hogarth, R., & Gaba, A. (2009). *Dance with Chance: Making Luck Work for You*, (Oxford: Oneworld Publications).

13 For example see Petropoulos, F., Makridakis, S., Assimakopoulos, V., & Nikolopoulos, K. (2014). '"Horses for Courses" in demand forecasting', *European Journal of Operational Research*, 237, pp. 152–63. Also see Syntetos, A. A., Boylan, J. E., & Croston, J. D. (2005). 'On the categorization of demand patterns', *Journal of the Operational Research Society*, 56, pp. 495–503.

14 Tetlock, P., & Gardner, D. (2015). *Superforecasting: The Art and Science of Prediction*, (London: Random House).

15 Armstrong, J. S., Green, K. C., & Soon, W. (2008). 'Polar bear population forecasts: A public-policy forecasting audit', *Interfaces*, 38 (5), pp. 382–405.

16 Armstrong, J. S. (ed.) (2001). *Principles of Forecasting: A Handbook for Researchers and Practitioners*, (Norwell, MA: Kluwer Academic Publishers). See also www.forecastingprinciples.com.

## Chapter 11: Knowing we don't know and the danger of knowing

1   Lovelock, J. (2014). *A Rough Ride to the Future*, (London: Penguin Random House), p. 69.

2   http://planetearth.nerc.ac.uk/news/story.aspx?id=1425

3   Moore, D. A. (2012). 'Stop being deceived by interviews when you're hiring', Forbes Leadership Forum, 7 February.

4   Wheelwright, V. (2012). 'It's your future... Make it a good one!', Personal Futures Network.

5   More detailed accounts of how to use scenario planning can be found in Wright, G., & Cairns, G. (2011). *Scenario Thinking: Practical Approaches to the Future*, (Basingstoke: Palgrave Macmillan) and in Ringland, G. (1998). *Scenario Planning: Managing the Future*, (Chichester: Wiley).

6   Based loosely on Goodwin, P. (2013). 'Combining scenario planning with multi-attribute decision making', in *Wiley Encyclopedia of Operations Research and Management Science*, (New York: Wiley).

7   Wilkinson, A., & Kupers, R. (2013). 'Living in the Futures', *Harvard Business Review*, 91(5), pp. 118–27.

8   Riggins, W. G. (2006). 'A perspective on regulatory risk in the electric industry', in Leggio, K. L., Bodde, D. L., & Taylor, M. L. (eds.) *Managing Enterprise Risk: What the Electric Industry Experience Implies for Contemporary Business*, (Amsterdam: Elsevier).

9   Wack, P. (1985). 'Scenarios: uncharted waters ahead', *Harvard Business Review*, September 63(5), pp. 72–89.

10  Ringland, G. (2006). 'Introduction to Scenario Planning', in Ringland. G., & Young, L. (eds.) *Scenarios in Marketing: From Vision to Decision*, (Chichester: Wiley).

11  Ibid.

12  http://www.bain.com/publications/articles/Management-tools-trends-2011.aspx

13  Hodgkinson, G. P., & Wright, G. (2002). 'Confronting strategic inertia in a top management team: learning from failure', *Organization Studies*, 23 (6), pp. 949–77.

14  This issue is discussed in more detail in Derbyshire, J., & Wright, G. (2014). 'Preparing for the future: development of an "antifragile" methodology that complements scenario planning by omitting causation', *Technological Forecasting & Social Change*, 82, pp. 215–25.

15  Taleb, N. N. (2007). *The Black Swan*, (London: Allen Lane).

16  Based on an example in Goodwin, P., & Wright, G. (2014). *Decision Analysis for Management Judgment*, 5th edition, (Chichester: Wiley).

17  Tversky, A., & Kahneman, D. (1983). 'Extensional versus intuitive reasoning: the conjunction fallacy in probability judgment', *Psychological Review*, 90 (4), p. 293.

18  Appleyard, B., 'Nassim Nicholas Taleb: the prophet of boom and doom', *The Times*, 1 June 2008.

19  Taleb, N. N. (2012). *Antifragile*, (London: Allen Lane).

20  Derbyshire, J., & Wright, G. (2014). 'Preparing for the future: development of an "antifragile" methodology that complements scenario planning by omitting causation', *Technological Forecasting & Social Change*, 82, pp. 215–25.

21  Ibid.

22  Lindblom, C. E. (1959). 'The science of "muddling through"'. *Public Administration Review*, 19, pp. 79–88.

23    Oster, E., Shoulson, I., & Dorsey, E. R. (2013). 'Optimal Expectations and Limited Medical Testing: Evidence from Huntington's Disease', *American Economic Review*, 103 (2), pp. 804–30.

24    'Predictive policing: Don't even think about it', *The Economist*, 20 July 2013.

25    Ferguson, A. G. (2013). 'Predictive Policing and Reasonable Suspicion', *Emory Law Journal*, 62, p. 259.

26    Steiner, C. (2013). 'Pop goes the algorithm', *The Futurist*, 47, p. 20.

27    *Lessons from the Human World of Hua Nan Zi* compiled by Liu An in the West Han Dynasty.

**Chapter 12: Conclusions**

1     Alfred Hickling, 'Is Liverpool really the centre of the creative universe? Alfred Hickling sets off on a Mersey odyssey to find out', *The Guardian*, 21 February 2007.

2     Taleb, N. N. (2007). *The Black Swan*, (London: Allen Lane).

3     'The Value of Money and Climate Data', Economics and Statistics Administration, United States Department of Commerce, 2 September 2014.

4     Grey, M. 'Public Weather Service Value for Money Review', Public Weather Service Customer Group Secretariat, March 2015.

5     Lee, K. K., & Lee, J. W. (2007). 'The economic value of weather forecasts for decision-making problems in the profit/loss situation', *Meteorological Applications*, 14, pp. 455–63.

6     Pappenberger, F., Cloke, H. L., Parker, D. J., Wetterhall, F., Richardson, D. S., & Thielen, J. (2015). 'The monetary benefit of early flood warnings in Europe', *Environmental Science & Policy*, 51, pp. 278–91.

7     Cohen, T., 'Here is the shopping forecast: How supermarkets use weather predictions to decide what to stock', *Daily Mail*, 16 August 2011.

8     http://www.bbc.co.uk/news/science-environment-24286258

9     Clegg, N., *Financial Times*, 24 June 2016.

# ACKNOWLEDGEMENTS

I am grateful to many people who, directly or indirectly, have helped during the preparation of this book. My wife, Chris, was never failing in her encouragement and patience, and she happily took on the tedious task of proofreading the original manuscript. Hannah Morgans read several sections of the book and provided helpful suggestions. I have also benefited from the expert advice of Sanjida O'Connell and Julian Baggini.

Special thanks go to my agent, Peter Buckman of the Ampersand Agency, for his excellent guidance and support. Thank you also to the team at Biteback Publishing, including James Stephens, Olivia Beattie, Ashley Biles and Sam Jones. Bernadette Marron's editing improved the manuscript significantly and I was hugely impressed with her attention to detail.

Finally, I must acknowledge my research collaborators with whom I have enjoyed exploring the fascinating topics of forecasting and scenario planning over many years. These include: Robert Fildes, George Wright, Richard Lawton, Michael Lawrence, Dilek Önkal, Andrew Pollock, Marcus O'Connor, Kostas Nikolopoulos, Wing Yee Lee, Aris Syntetos, John Boylan, George Cairns, Sinan Gönül, Mary Thomson, Melanie Kreye, Sheik Meeran, Karima Dyussekeneva, Shanshan Lin, Haiyan Song, Maryam Mohammadipour, Baris Yalabik, Fotios Petropoulos, Esra Öz, Semco Jahanbin, Joao Quariguasi Frota Neto, Valeria Belvedere, Xia Meng and Rob Hyndman.

# INDEX

**320PP PAPERBACK, £9.99**

2016 marked the birth of the post-truth era. Sophistry and spin have coloured politics since the dawn of time, but two shock events – the Brexit vote and Donald Trump's elevation to US President – heralded a departure into murkier territory.

From Trump denying video evidence of his own words, to the infamous Leave claims of £350 million for the NHS, politics has rarely seen so many stretching the truth with such impunity.

Bullshit gets you noticed. Bullshit makes you rich. Bullshit can even pave your way to the Oval Office.

This is bigger than fake news and bigger than social media. It's about the slow rise of a political, media and online infrastructure that has devalued truth.

This is the story of bullshit: what's being spread, who's spreading it, why it works – and what we can do to tackle it.

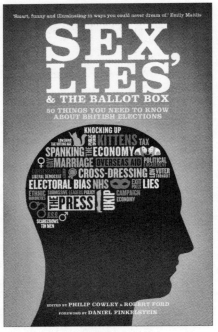

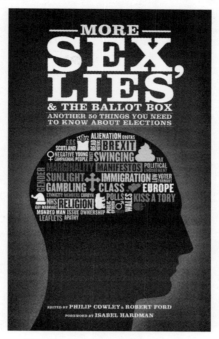